Sustainable Rural Technologies

Sustainable Rural Technologies

– by –

M.S. Virdi

Ex-Director

CSIR Polytechnolgy

Transfer Centre, Bhopal

Patron Satpura Vigyan Sabha

Bhopal

2018

Daya Publishing House®

A Division of

Astral International Pvt. Ltd.

New Delhi - 110 002

First Published, 2009
Reprinted, 2018

ISBN: 9789351242000 (Hardbound)

Publisher's note:

Every possible effort has been made to ensure that the information contained in this book is accurate at the time of going to press, and the publisher and author cannot accept responsibility for any errors or omissions, however caused. No responsibility for loss or damage occasioned to any person acting, or refraining from action, as a result of the material in this publication can be accepted by the editor, the publisher or the author. The Publisher is not associated with any product or vendor mentioned in the book. The contents of this work are intended to further general scientific research, understanding and discussion only. Readers should consult with a specialist where appropriate.

Every effort has been made to trace the owners of copyright material used in this book, if any. The author and the publisher will be grateful for any omission brought to their notice for acknowledgement in the future editions of the book.

Published by : **Daya Publishing House®**
A Division of
Astral International Pvt. Ltd.
– ISO 9001:2015 Certified Company –
4736/23, Ansari Road, Darya Ganj
New Delhi-110 002
Ph. 011-43549197, 23278134
E-mail: info@astralint.com
Website: www.astralint.com

Digitally Printed at : **Replika Press Pvt. Ltd.**

To my inspiration

***Harsharan Virdi** (Wife)*

Acknowledgements

1. Central Building Research Institute (CBRI) Roorkee–247667, Uttarakhand.
2. Central Food Technological Research Institute (CFTRI) Mysore–570 020, Karnataka.
3. Central Leather Research Institute (CLRI) Adyar, Chennai–600020, Tamil Nadu.
4. Central Mechanical Engineering Research Institute (CMERI) MG Avenue, Durgapur–713209, West Bengal.
5. Mechanical Engineering Research and Development Organisation (MERADO), Gill Road, Ludhiana (Punjab).
6. Central Scientific Institute Organisation (CSIO) Sector 30-C, Chandigarh–160030 (UT)
7. Central Institute for Post Harvest Engineering and Technology (CIPHET) PAU Campus, Ludhiana, (Punjab).
8. Central Salt and Maire Chemical Research Institute (CSMCRI), Gijubhai Badheka Marg, Bhawnagar–364 002, Gujrat.
9. Institute of Himalayan Bioresource Technology (IHBT) Palampur–176661.

10. Central Institute of Medicinal and Aromatic Plants (CIMAP) P.O. CIMAP, Near Kukrail Picnic Spot, Lucknow–226015.
11. Indian Institute of Petroleum (IIP). P.O. IIP, Mohkampur, Dehradun-248005, Uttarakhand.
12. National Botanical Research Institute (NBRI) Rana Pratap Marg, Lucknow–226001.
13. National Environmental Engineering Research Institute (NEERI) Nehru Marg, Nagpur-440020, Maharashtra.
14. AMM Muragappa Chettiar Research Center, Taramani Chennai.
15. Advance Materials and Processes Research Institute (AMPRI earlier RRL), Hoshangabad Road, Near Habibganj Naka, Bhopal-462026.
16. Institute of Minerals and Metals Technology (IMMT, earlier RRL) Bhubaneswar–751013, Orissa.
17. Indian Institute of Integrative Medicine (IIIM earlier RRL) Canal Road, Jammu–181001 (JK).
18. North East Institute of Science and Technology (NEIST, earlier RRL) Jorhat–785006, Assam.
19. National Mushroom Research Centre, Solan H.P.
20. Central Institute of Agricultural Engineering (CIAE) Nalu Bagh, Berasia Road, Bhopal.
21. Indian Resins and Gums Research Institute (Earlier ILRI) Namkum Ranchi, Jharkhand.
22. Centre of Science for Villages (CSV), Wardha, Maharashtra.
23. Appropriate Rural Technology, Institute (ARTI), Phaltan Dist. Satara, Maharashtra.
24. Indian Avian Research Institute Izzat Nagar–243122 Dist. Bareli, UP.
25. Sheep and Wool Research Institute, Avika Nagar, Rajasthan and Kulu H.P.
26. Central Institute of Fisheries Education, Versova, Mumbai.
27. Central Glass and Ceramic Research Institute, (GCRI) Jadavpur, Kolkata.

28. Development Alternatives (DA), Taragram Orchha, UP and New Delhi.
29. Central Ground Water Board, Bhopal and New Delhi.
30. Department of Renewable Energy Resources, GOI, Bhopal.

Foreword

With more than 70 per cent population of the country continuing to reside in the rural areas, it should be the endeavor of everyone concerned to see that the farmers and rural artisans get best returns for their investment and labour. Accelerated rural development is also necessary to check migration from rural to urban areas. This, in turn, requires that productive employment opportunities be available in rural areas throughout the year. A lot of work is being done in the Agricultural Universities and Research Institutions not only to increase .productivity but also for modernizing agricultural equipments to reduce the drudgery of the agricultural labour.

Efforts are also being made for value addition in agriculture through agro-processing. Minor forest produce, with proper organizational and marketing arrangements, can not only provide livelihood to the rural people, including the tribals, but also keep them engaged during the lean period between the agricultural seasons. . Research has also been done, for construction of low cost housing which is, at the same time, eco-friendly and environmentally suitable houses. Shri M.S. Virdi has done a sincere and commendable effort to compile the various R&D efforts of the Research Institutions in these fields. The document is certainly going to be of help to those who are committed to the cause of rural development.

Raghuvansh Prasad Singh
Minister of Rural Development
Government of India,
Krishi Bhawan, New Delhi – 110 114

Contents

Introduction

More than 75 per cent country's population lives in villages and tribal areas. During last two decades there has been rapid migration from rural areas to cities and urban habitats to survive and earn livelihood which gives rise to environmental problems in urban areas as most of these migrants lives in Slums. It has been observed that there is idle time between sowing and harvesting which leads to serious problems. Creating job opportunities in rural areas may help to prevent migration, mitigate serious problems and utilize rural resources. It is believed that if entire agriculture produce is processed in rural areas it may offer jobs in millions and improve economy of villages.

Cultivation of medicinal and aromatic plants will certainly bring higher profits. Cultivation of mentha in U.P. has created employment for one million people.

Food processing offer immense potential for value addition of horticulture produce. At present only 1 per cent of country's produce is processed. Setting up of food processing units in rural areas and their linkage with large units can give boost to Horticulture. Livelihood of Tribal population depend upon Minor Forest Produce

(MFP) which is available at different time periods of the year. A calendar could drawn for collection of MFP and Centers for preservation, processing and marketing of MFP would improve economy of tribal population. Infact, substantial portion of MFP is not even collected.

Low cost housing using locally available materials, potable drinking water, sanitation is basic need of rural areas.

Some employment generating activities should also be taken up under Employee Guarantee Schemes of Central and State Governments.

Selection of appropriate technology for sustainable rural development is a vital issue and one must keep in mind the local resources, type of man power and marketability, while taking decision. Author, during last three decades introduced several new technologies in Central India and results have been very encouraging. Although considerable RandD has been carried out by National Laboratories of CSIR, ICAR, NGOs etc. but these have not reached the remote areas of country where these could he utilized. The book will be useful for Departments of Rural Development and Panchayats, District Industries Centers, Community Polytechnics, NGOs and entrepreneurs, individuals etc.

More 100 proven technology have been covered in the book. In some cases estimated cost of Investment etc has been given as guideline it may vary from place to place.

Author is grateful to Directors and Scientists of RandD institutes where technology have been developed which have been described in the book.

Author expresses sincere thanks to Shri A.C. Chandra, Dr. B.N. Johri and Dr. Anil Prakash for their valuable suggestions and Mr. Younus Khan for typing the manuscript.

M.S. Virdi

Chapter 1

Building Materials and Components

Micro-concrete Roofing (MCR) Tiles

Housing is the basic need of every human being. Central Government as well as State Governments are keen to provide housing to everyone and especially to weaker sections of the society. The cost of housing construction is increasing day by day due to rising cost of building materials. The Government sponsored agencies and individuals are looking towards low-cost building materials without compromising with the quality. Research and development organizations of the country have developed low-cost building components for walling and roofing which enable one to build improved houses at comparatively cheaper cost. Price of corrugated G.I. sheets used for roofing, is continually increasing while its use doesn't provide thermal comfort in a tropical country like India. Use of Asbestos–Cement sheets is being discouraged due to health hazards associated with asbestos fibre. MCR (micro-concrete roofing) technology meets the growing demand for high quality roofing. MCR tiles are cost effective and extremely versatile roofing material. MCR tiles can be used on steel and wooden under-structure to make attractive roofs on residences, farmhouses, verandahs and pavilions. In areas with heavy rainfall, MCR tiles are used extensively as cladding material offering both water proofing and aesthetic appeal. It has been used extensively in cost effective housing schemes, workplaces, restaurants and poultry farms. MCR tiles are ideally suited to replace thatch and fire clay tiles in rural areas. The MCR technology in India has been promoted by M/s Development Alternatives and TARA.

Advantages

Traditional roofing materials have their own limitations but MCR tiles are versatile, economical, aesthetic, eco-friendly and are made using minimum energy. MCR tiles provide better thermal comfort over asbestos and GI sheets.

Other advantages are:

1. It is leak proof
2. It is fire proof
3. It can be made in any colour
4. It is eco-friendly and doesn't contain asbestos fibre and hence good for health.
5. Production process is pollution free
6. Utilises waste (stonedust)

Process of Manufacturing

Step–I

Selection of raw material

Cement (43 Grade)

Stone dust (Fine)

Aggregate (<6mm)

Water

Step–II

Mix preparation: The above raw materials are mixed in the following ratio:

Cement	:	Stone dust	:	Aggregate	:	Water	
1	:	1.5	:	1.5	:	0.5	(by weight)

Step–III

Mix Vibration: The prepared mix is put on a polythene sheet and placed over vibration table for about 45 sec. A nail is inserted in the rib for holes. It is then aligned on the alignment tool.

Step–IV

Moulding and stock curing: The polythene sheet is then carefully put on the plastic moulds for getting a uniform shape. The moulds are then stocked one above the other in a set of 25 for next 24 hr.

Step–V

Demoulding and water curing: The tiles are then demoulded and separated from the plastic sheets. The nails are also taken out and edges sharpened for better finishing. Then the tiles are kept in the curing tank for another 7 days.

Step–VI

Strength and quality testing: After the tiles are taken out from the tank these are stored for next 21 days in open or shed. Tiles from the stock are taken at random for testing in accordance with standard specifications. Then the tiles are ready for sale.

Product Specifications

Clear Length	488 mm
Length after overlap	400 mm
Clear width	240mm
Width after overlap	200mm
Thickness	8mm/10mm
Corrugation depth	50mm
Normal weight	2.25-2.60 kg
Purlin spacing	400mm
Load bearing capacity	80/100 kg

Operating Parameters

- Capacity: 200 tiles/day
- Machine/equipment:
  - TARA-MCR Machine: 1

- ☆ Moulds (PAN): 200
- ☆ Alignment tool: 200
- ☆ Bend test equipment: 1

☆ Power requirement: 5 units/day, single phase 1 KW electric connection would be sufficient.

☆ Other requirement: water 100 l./day

☆ Raw material

- ☆ Cement: 2 bags/day
- ☆ Sand/stone dust: 4 qt./day
- ☆ Aggregate: 4 qt./day

☆ Packaging: No special packaging is required.

☆ Manpower requirement for production: 4

Capital Cost

(*a*) Land and Building with auxiliaris, 300 Sq.ft.: Rs. 50,000

(*b*) Machinery and Equipment: Rs. 75,000

(*c*) Working capital: Rs. 2,00,000

(*d*) % Profit: 50 per cent

Selling price per tile: Rs. 6.00

Water and Fire Proof Mud Walls and Thatch

In the rural areas mud walls get eroded due to rainfall and the walls collapse in a short time of 2-3 years. CBRI Roorkee has developed an economical and effective method to construct non-erodable mud walls which have a life of 8-10 years. Non-erodable mud walls are prepared by mixing bitumen cut back with specified quantity of mud mortar. Bitumen cut back and kerosene are mixed in the ratio of 5:1 by weight. For example 55 kg of bitumen is heated till melting and poured into 11 l. of kerosene and continuously stirred. This bitumen is poured into 1.0 cum. mud mortar and thoroughly mixed by overturning with sickle and puddle, which is sufficient for 10 sq.m. of wall surface.

Dressing of Wall Surface

☆ Fill all cracks

☆ Make the mud wall surface uniform and level.

- ☆ Scrap undulations and irregularities
- ☆ Make wall moist by sprinkling water
- ☆ Start plastering from top to bottom and keep plaster thickness–12mm
- ☆ Allow plaster to dry for a day.
- ☆ Fill the cracks developed during drying to get crackfree surface.
- ☆ Apply two coats of gobar
- ☆ White wash all walls.

For making the thatch water and fire proof mud plaster is applied in two layers on the top of panels, the lower layer consists of traditional mud mortar while the upper layer consists of 12mm non-erodable mud plaster prepared as above. Two coats of bitumenized gobar paste prepared by mixing cow dung and mud in ratio of 1:1 and adding bitumen cut back in the specified proportion are applied over dried-non-erodable mud plaster surface to fill up cracks developed while drying the plaster. About a litre of kerosene oil and 22 kg. of bitumen are required to cover 100 sq. ft. (9.3 sq.m) of roof area.

Several demonstration-cum-training programmes have been conducted to popularize this technology which is most appropriate for rural areas.

Red Mud/Rice Husk Cement

Cement is major raw material used in housing industry and it is not possible to construct a durable house without cement. It is used in joining of bricks, for plastering and RCC/RBC roofing etc. Red-mud which is a waste of aluminium industry could be used in the manufacture of cement substitute which is cheaper and has similar properties. The price of cement is rising continuously. Therefore, it is high time to find dependable substitutes.

In the process of manufacture. red-mud, rice husk and coarse sand (murram) are mixed in prescribed ratio and fired in a kiln. The ash so obtained is ground with slaked lime using pulverizers. The pulverized material is then weighed and packed in bags for supply in the market.

In the process for production of rice husk cement, rice husk, lime and clay are mixed in the ratio of 50:25:25 by weight and slurry is made by using water. Subsequently, nodules/round balls are made and sun dried which are later fired in a trench kiln. No extra fuel is required since rice husk is a self burning material. The balls are micro-pulverised for use as cement substitute. The rice husk cement could be used for joinery, plastering, foundations etc. but not for RCC roofs.

A 5T/day capacity unit could be set up with an investment of Rs.5.00 lakh.

Red mud is available as waste from aluminium plants and rice husk is available from rice mills while lime is available from local/nearby markets.

Interested entrepreneurs can contact AMPRI (RRL), Bhopal for technical guidance.

Fly Ash Bricks

Fly ash emitted by thermal power stations is one of the highest environmental Hazards and Government of India has set up a National Fly Ash Mission for utilisation of Fly ash. RandD has been carried out for its use in agriculture as a fertiliser, aggregate for road making and fly ash-polymer composites for door panels, building bricks, construction of dams, pozzolanic cement etc. In the traditional process of brick making agricultural soil is being used and thus deprives land from agricultural use. In India there is need of about 1.5 crore houses in rural areas which require crores of bricks. The bricks using fly ash have very high compressive strength and one can construct three-storied house without use of columns. There are two processes of making bricks using fly ash:

Clay-fly Ash Bricks

In this process 30-35 per cent fly ash is thoroughly mixed with clay and bricks are cast by hand moulding process and plate vibrator or by using moulding press. The moulded bricks are first sun dried then fired in brick kilns to a temperature of 950 ±20°C. Thus, red bricks are made having 40-50 kg/cm^2 strength and it requires 25 per cent less fuel during firing.

Sand-lime–Fly Ash Bricks

In this process fly ash, sand and lime are mixed in the ratio of 80:10:10 and the mix is pressed in the moulding press. The moulded bricks are sun-dried for 2-3 days and cured in water for 28 days or steam cured for 4 hours. These bricks have sharp edges and strength of 40-50 kg/cm^2 and thus do not require plastering.

CBRI Roorkee has developed a vibro press in which 4 bricks are made in one operation or 3000 bricks/shift/day. The cost of machine is Rs.70-80,000 and it is now commercially available.

For technology, entrepreneur can contact CBRI Roorkee, and AMRI (RRL), Bhopal.

Ferro-cement Water Tanks and Other Products

In the Ferro-cement techniques high strength galvanised wire–mesh is sand witched between layers of cement and sand mortar in the ratio of 1:2.

Ferro-cement water tanks have replaced M.S. and concrete tanks due to cheaper cost, lightweight and long life requiring almost zero maintenance. There is good demand of water tanks in urban as well as rural areas. CSIR laboratories have transferred this technology to more than 150 parties and ferro-cement tanks are in use in various parts of the country.

In a semi-mechanised process wire mesh is wound tightly around cylindrical mould and cement sand mortar is applied in such a way that wire mesh is sandwiched between the layers of mortar so as to achieve wall thickness of 12mm for tanks upto 500 l capacity. Base and cover are cast separately. Base is joined to the cylindrical surface. The tanks are water cured for 15-20 days by sprinkling water or immersing in water. The inside surface of the tank is coated with anti-fungal/anti-algal food grade resin. Outer surface is painted with silver paint to reflect heat.

Water tanks upto 10,000 capacity could be made by casting the cylindrical surface in four segments, which are joined together with base *in situ*. A manufacturing unit for making 10 tanks/day of capacity upto 500 liters requires investment of Rs. 2.00 lakh.

The following products also can be manufactured in the same production facility with minor addition of moulds.

- ✰ Ferro-cement grain storage silos/bins.
- ✰ Ferro-cement door shutters.
- ✰ Ferro-cement roofing sheets.
- ✰ Ferro-cement manhole covers.
- ✰ Ferro-cement irrigation and drainage channels for rain water harvesting.
- ✰ Ferro-cement earth quake resistant house.
- ✰ Ferro-cement cupboards.
- ✰ Ferro-cement septic tanks.
- ✰ Ferro-cement flower pots.
- ✰ Ferro-cement animal feed units etc.

Entrepreneurs interested in technology transfer and training may contact CBRI, Roorkee, SERC, Chennai as well as the author of this book.

Stone Masonry Cement Blocks

In areas where bricks are costly and of poor quality it is appropriate to use solid cement concrete blocks for constructing walls. In Andaman and Nicobar islands this technology is very popular due to non- availability of bricks. The blocks could be cast in any size. CBRI has developed lean concrete mixes using 50mm and below coarse aggregates and fine aggregates. Compressive strength of 70-100 kg/cm^2 has been achieved by using cement aggregate ratio of 1:18. In the process cement aggregate mix is filled in the moulds and placed over vibrator of 3000 VPM (0.5 kw). An experienced mason with three helpers can produce 400-500 blocks in a day. Raising the walls using these blocks is faster and walls do not require plastering. The blocks are cured for 3 weeks by frequent sprinkling of water followed by air drying for three weeks. A casting platform of 10m x 5m size is sufficient to produce 500 blocks/day.

Now a days egg laying type portable machines are commercially available having casting capacity of 1000 blocks of size 290 x 190 x 140mm. in a shift of 8 hr. It is now possible to make hollow blocks, pavement blocks, tree-guard sections etc. using available vibrating presses.

Stabilised Mud Blocks

In rural areas bricks are cast in wooden moulds, sun dried and fired in open 'Bhatta' using wood as fuel which results in depletion of forests. CBRI, Roorkee has developed stabilized mud blocks/bricks which do not require firing and have compressive strength of 25-30 kg/cm^2. In the production process, mud is prepared by usual methods and 5-7 per cent hydrated lime or 2-3 per cent cement by weight is added to the mud and thoroughly mixed using showels before pressing in the Moulding Press. The blocks are subsequently dried and used in brick masonry. An investment of Rs.2.00 lakh is required to produce 1200 blocks/day.

Precast Stone Masonry Blocks

In many parts of the country stone is abundantly available and walls are made in the form of random rubble masonry and wall thickness is generally between 38 to 45 cm. Alternatively, the stones are dressed which require more skilled labour and time and the average thickness of the plaster is 20-25 mm because of unevenness of stones. Central Building Research Institute (CBRI) has developed blocks for reducing the thickness and level of skill required in random rubble masonry walling. In this process stone rubble is filled in iron/wooden moulds and a mix comprising of cement, sand and fine aggregate in ratio of 1:6:8 is poured into the mould which is later kept on plate vibrator so that the cavities are filled with the concrete mix. Subsequently, the blocks are sun dried and water-cured as usual. The scheme has been successfully adopted for load bearing and non-load bearing construction upto three storeys by various organisations in different parts of the country and other Asian and African countries.

Preservation of Mud Walls

Generally mud walls are not stable and get eroded by rain water. It becomes necessary to enhance their life. CBRI Roorkee has developed techniques to protect mud walls. In this process the wall is cut 2-3 cm deep just below ground and wall is straightened by cutting extra mud upto a height of 60-75 cm. A trench of 23 x 23cm is dug around the mud wall. The foundation is filled by a mixture of gravel and mud and grouted by manual process. A brick wall of 115 cm. thickness is raised close to the mud wall upto a height of 60 to 75 cm. and on top of it a full brick (23 cm) is fixed with mortar. Extra 12.5cm portion of top brick is fixed with mud wall by scratching it. The brick wall is lined with cement: sand mortar in the ratio of 1:4.

This technique protects the mud walls around the mud houses.

Chapter 2

Agricultural Implements

Central Institute of Agricultural Engineering (CIAE) Bhopal, a constituent of Indian Council of Agriculture Research (ICAR) has developed a range of agricultural equipment for modernization of small as well as large farms. These include manually operated, animal drawn power tiller, self propelled, tractor operated and power operated hand tools, implements, equipment and machinery. Prominent among these are: blade harrow, multipurpose tool frame, till planter, auger digger, seed-cum-fertiliser drill, no-till drill, inclined plate planter, pneumatic planter, plot drill, rice transplanter, vegetable transplanter, hand weeders, power weeders, sprayers, tiller operated cultivator, serrated sickle, reaper windrover (walk behind), tractor/ power tiller operated vertical conveyor reapers, multi-crop thresher (5 hp) and high capacity multicrop spike-tooth thresher (20 hp) for cereals, millets, oil seeds and pulses. In the field of agro-processing Institute has developed cleaners, graders to improve quality of grains, maize shellers, ground nut decorticator, grain pearlers, mini grain mill and mini-dall mill. For details one should contact Head,

Technology Transfer Division, C.I.A.E. Nabi Bagh, Berasia Road, Bhopal-462038, M.P. The Institute also provides designs of these machines on nominal terms.

Regional Research Laboratory (RRL), Bhubaneswar has developed a simple machine for separation of straw from paddy and to subsequently clean the paddy for supply to rice mills. It is operated by one h.p. motor and processes 300 kg of paddy per hr.

All these machines could be manufactured by setting up a fabrication shop having equipment/machinery like lathe, drilling machine, cutting saw, welding set etc. Investment of Rs. 2.00–3.00 lakhs on machinery would be sufficient to set up manufacturing facilities.

1. Naveen Sickle

The device follows predominantly the principle of friction and shearing with serrated blade to cut plants. The cutting force is drastically reduced compared to conventional sickle which uses impact principle.

It consists of serrated blade, ferrule and wooden handle. Appropriate size of raised wooden handle protects the farmer/ operator. The CIAE has also developed serrating machine for this purpose which costs around Rs. 30,000 and takes 3-5 minutes to serrate a sickle.

2. Animal Drawn Digger

It is operated by a pair of bullocks for digging of groundnut and potato in addition to secondary tillage operation like harrowing. The unit consists of beam, frame, handle, ground wheel depth adjustment mechanism and V-blade. It takes 8-20 hr per hectare and price is around Rs. 3500.

3. Grubber Weeder

It is a simple and light weight, manually operated tool for weeding and interculture in upland row crop in black cotton soil. It consists of long handle, ferrule, three tynes and sweep type blades. The operator uses pull force to break the soil crust and uproots the reeds.The cost of the device is Rs.150-200. Animal driven weeders are also available.

Manually operated Peg Type dryland weeders have also been developed for weeding and interculture in crops like ground nut, sunflower, soyabean and sorghum (Price Rs.400 approx.).

4. Self Propelled Reaper

It is operated by 6 hp diesel engine. Crop cutting and conveying mechanism is mounted on the front of the unit. It consists of handle, engine, cutter bar, crop row divider (4 nos), star wheel, conveyor belt and bugged wheel. The unit is suitable to harvest cereals and oil seeds. It requires 5 man hr/ha. and cost of the reaper with prime mover is Rs. 60,000–70,000.

5. Seed Drill

CIAE has developed single row manually operated seed drill, animal drawn two row seed drill and two row seed cum fertiliser drill. The cost ranges between Rs. 1500-3500 for different types.

Two row animal drawn drill is simple, light weight and compact machine to sow crops like wheat, gram, sorghum, soyabean, lentil, pea, sunflower, safflower etc. in black cotton soil under rain-fed conditions. The fluted roller metering mechanism fitted in the unit gets the drive from ground drive wheel of 300 mm dia through chain and sprocket. The shoe type furrow opener with non-closing boot places the seed at desired depth.

6. Improved Bakhar

Improved bakhar is suitable for seed bed preparation in black soil. This has a modified blade replacing conventional curved blade for reduced draft. The roller provided at the rear crushes the soil clods. The unit consists of main frame, V-blade, handle, roller and beam for adjustment of blade according to soil.

7. Manual Rice Transplanter

It is a six row manually operated equipment to transplant mat type rice seedling in rows in puddled soil. It consists of handle, picker assembly, seedling tray, tray indexing mechanism, main frame, base frame, seedling pushing lever, feeder link assembly, wooder skid and pivot arm assembly. It requires 40-50 man hr for planting one hectare and costs around Rs. 6,500-7,000.

8. Multi Crop Thresher

Multi Crop Thresher has also bean developed suitable to thresh wheat, maize, rice, grain pigeon pea, soybean, mustard, safflower, linseed etc.

Chapter 3

Food Processing

A large variety of fruits and vegetables grow under sub-tropical and temperate climatic conditions due to diverse agro-climatic conditions prevailing in different regions of the country. India produced annually 32 million tonnes of fruits which is 12 per cent of worlds' production and 71 million tonnes of vegetables which forms 15 per cent of world production. The major fruits grown are banana, mango, citrus (orange, mausami), guava, grapes, apple, pineapple etc. Although India is a very large producer of fruits per capita consumption is mere 75 gm. Further, 20 to 22 per cent of production of fruits is lost due to spoilage at various post harvest stages and transportation which amounts to Rs.3000 crores in a year. Nearly 76 per cent of production of fruits and vegetables is consumed fresh while losses account for 20–22 per cent. Only 2 per cent of vegetable production and 4 per cent of fruit production is being processed compared to Brazil–70 per cent, Malaysia-80 per cent, Philippines-78 per cent, and Thailand-30 per cent. There are 7000 processing units in organised sector and thousands in unorganised sector and the industry employs 1.6 million workers. The turnover of industry is Rs. 3,00,000–3,50,000 crores. The demand for processed fruits and vegetables comes from both domestic and export markets. A substantial share of production is consumed by the defence deprtment, hotels and restaurants. Household consumption accounts for less than 50 per cent production. Export of processed products is 13 per cent of total exports of the country. The bulk share in exports is of mango products. India's share in world trade in this sector is around one per cent.

The average Indian farmer is poor, high quality planting material is not easily available and he is also not fully aware of modern agro-economic practices. The yield per acre is quite poor compared to other Asian countries.

The existence of long marketing channels with a number of middlemen results in low return to the growers. Direct linkage between growers and processors is most essential.

The storage and transport infrastructure is highly inadequate resulting in high post-harvest wastage and losses. The average Indian farmer is hesitant to take up horticulture requiring high investment, long gestation and uncertain returns.

The major constraints for processors are:

1. Non-availability of high/uniform quality of fruits and vegetables.
2. The domestic demand of processed foods is low due to availability of fresh fruits.
3. Absence of captive farming and corporate farming.

Constraint of consumer is high cost of processed foods, which should be brought down to suit the pocket of middle class consumer.

Fruit and Vegetables Preservation

Processing of Orange, Lemon and Mango Process Flow Sheet

Raw materials Procured from farm → Weigh → Wash → Deskin if required ↓

Hot blanch → Cut → Pulp or Juice → Pasteurize ↓

Quality testing → Hot fill → Seal

Scale of Operation

Minimum Production Capacity

Pickle	10T/year
Jam	15T/year
Mango Juice	01T/year
Tomato sauce/ketch-up	35T/year
Misc	5 T/year
	66 T/year

Machinery/Equipment–Rs. 3.50 lakhs.

1. Mini Boiler–100 lb/hr cap.
2. SS. pulping m/c (80kg/hr with ½ h.p. motor (BSB)
3. Solar drier-1
4. Electric sealing m/c 12" (power op)
5. Crown corking m/c
6. R.O. cap sealing m/c
7. Fruit/veg. slicer ½ hp heavy
8. Table with SS screw type juice extractor (BSB model 100)–½ hp moor
9. Table with S.S. top (3' x 8')
10. Cutting knives
11. Weighing balance–Single pan 1 g–2 kg.
12. Weighing Balance (100 kg.)
13. Food grade plastic containers (500 l–2nos., 100 l-10nos). 20 l.–10 nos.
14. S.S. hand peeling, slicing and cutting tools
15. Pulverizer 2 hp
16. Hand operated juice extractor Al body with S.S. screw
17. Lab. and quality control equipment: Refractometer, balance, glassware, reagents etc. and misc. accessories *e.g.* S.S. sink, worker's tools etc.

Economic Feasibility (in lakhs)

Building 2000 sq.ft.	4.00
Machinery	3.50
Pre-op./consultancy	1.00
	8.50
Working capital	6.00
Total	14.50
Sales	30.00
Cost of Prod.	24.00
Profit	6.00

$$\% \text{ Profit} \quad \frac{6.00}{14.50} = 45\%$$

Source of Technology

Consultancy and Training can be obtained from Central Food Technology Research Institute, (CFIRI) Mysore.

A small plant of this capacity is functioning at Kesla near Itarsi M.P. set up under UNDP Assistance. Setting up of such a mother plant has catalysed cultivators of fruits and vegetables in rural areas. These small plants can also supply semi-processed products like Juice/Pulp to large units apart from supply of packaged products in local market.

Soya Milk

It is a cheap and nutritious substitute for Buffalo/Cow milk. It has same protein and fat content but lacks lactose sugar found in cow's milk. It is possible to produce 8 l of milk from one kg of soya beans.

Process

Shown on next page.

Soya beans

↓

Soak in water for 6-8 hour

↓ Drain water

Blanch for 20 minutes with Baking powder → Reduces beany flavour

(← 2 Teaspoon of baking powder for 1 kg of soybean)

↓ Cool

Dehull → Hulls

↓

Grind dehulled kernels

↓

Centrifuge in mixie with 8 water

↓ Filter → Ochhra → Buddees

Raw soya milk

↓

Sterilise/Boil

↓

Add sugar/flabour

Economics

1. Capacity — 400 L of milk (processing of 50 kg of beans)
2. Premises (rental) — 2000/pm
3. Machinery:
   - (*a*) Containers for soaking and working
   - (*b*) De-huller manual
   - (*c*) Wet grinder
   - (*d*) Centrifuge

   } 1.00 lakh

4.	Working Capital:	0.25 lakh
	Total project cost	1.25 lakh
	Sales 400 L @ 5/-	2000
	Production cost @ 3/-	1200
	Profit	800/per day.

Edible Soyabean Flour

Soyabean is consumed in China, East Asian countries as a traditional food from time immemorial. Soyabean is a rich and cheap source of protein and excellent raw material for fabricated food products with desired composition, texture and flavour. Malnutrition prevailing in most of the rural area of the country could be tackled by the use of soya foods. Five per cent soya flour mixed with chapati flour and 15-20 per cent in occasional ready to eat items not only improves taste but also enhances the protein and mineral contents of food.

Soyabean has 5-17 per cent moisture, 36-50 per cent protein (average 40 per cent) 14-24 per cent carbohydrates, 13-24 per cent fat and 3-6 per cent ash.

The process for the production can be adopted on cottage scale. Easiest process to manufacture soya flour is first to dehusk beans, rinse in water for 3-4 hr and cook the beans for 10 minutes to reduce the beany flavour. Subsequently, water is drained off and beans are spread on the floor for sun drying or else it may be dried in electric hot air drier or solar drier. Slight roasting may enhance shelflife of flour. The soya seeds are then ground to appropriate particle size. It is advised to mix 500gm of soya flour in 10 kg of wheat flour.

Approximate composition of soyabean flour is:

Protein	50-55 per cent
Carbohydrate	30 per cent
Fat	0.6 per cent
Moisture	7 per cent
Fiber	2.7 per cent
Ash	6 per cent

Equipment

	Rs.
Huller (CIAE design)	15,000
Roaster	15,000
Disc mill grinder	20,000
	50,000

Leaf Cups and Plates

Leaf cups and Plates are used in social and religious functions for serving food stuffs. These are made from leaves of Banana, *Butea mono-sperma*, Bauhinia, Sal and sheaths of Arecanut palm using traditional methods which have inherent disadvantages like holes and uneven shape. CFTRI has developed a pedal operated as well as a hand operated leaf cup machine which make attractive leaf cups and plates having dimensional stability, attractive appearances, shape, finish and hygienic quality. These products are now available throughout the country and is perhaps one of the most popular technologies.

The salient features of the machine are:

(*a*) It is manually operated and all operations of folding, trimming, pressing into shape and drying are done in a single step by pressing the lever.

(*b*) The easily available 300 watt electric filaments are used which work on domestic electric line supply.

(*c*) 250-300 cups or 100 plates could be made in an hour by a skilled operator and a helper.

(*d*) Since leaf cup is subjected to heating in the range of 150-200° C for about 10 seconds these get sterilized and do not require any antifungal treatment. Machine made cups/plates packed in sealed PVC bags are bacteria free.

(*e*) The machine can also be used for making donas/plates from laminated card sheets or aluminium foil.

In the process of manufacturing, green leaves are well dried and then soaked in water for 5 minutes and wiped with a clean cloth to remove dirt/dust and again dried to retain their pliability. The leaf is kept in the dye having heater in the male (upper) portion and in single operation it adopts the shape of the dye and its edges are nicely trimmed.

The cost of the machine with extra dyes is within Rs.10,000.

Production capacity: 1500 cups/500-600 plates/day

One operator and helper can earn: Rs. 100/day.

Some Manufacturers

1. A.M.I. Engineering, Opp. Veena Cinema, Station Road, Patna-1
2. Nalanda Agro Works, Ind. Est. Nalanda Nagar, Kurji, Patna-800 020
3. Kalpataru Enterprises, Ind. Area, Amausi, Lucknow.
4. Bharat Engineering Works, Near Budhwaria Bus Stand, Ujjain.
5. Shah Engg. Co. C-18, Ajanta Commercial Centre, Ashram Road, Ahmedabad–380 004.

Vinegar

Vinegar production is a traditional rural industry and there is scope to improve the process of vinegar production. The raw materials like sugar cane, black berry (Jamun) and Sapota are available in abundance in rural areas. Vinegar is used for sprinkling over salaad, pickles and many food preparations. There is also market for vinegar in restuarants in cities. It is a profitable rural industry.

The process involves keeping the pulp and juice in wooden vats or large earthen pots with the addition of culture. The pots are kept for 15-20 days duly covered with Jute cloth to provide heat and accelerate fermentation. Thereafter, the material is filtered and sterilised by application of slow heat. The finished product is packed in glass bottles.

Economics

Capacity	5000 l/year
Vats and utensils	Rs.50,000
Working capital	Rs.25,000

Total investment	Rs. 75,000
Sales 5000 L x Rs. 20/- per l. =	1,00,000
Cost of Prod. =	50,000
Profit Rs. =	50,000

Source of Technology: CFTRI, Mysore or Traditional Industries.

Energy Food

There is a prevailing problem of malnutrition in rural and tribal areas of India due to immense poverty and the percentage of population suffering from mal-nutrition is increasing. It is essential to combat nutritional deficiency by providing cheaper supplementary food to pregnant and anaemic women and school children. It can also be served in drought affected areas.

Energy food is based on blends of locally available raw materials such as wheat, bengal gram, edible ground nut flour and Jaggery forti-fied with vitamins and minerals. Typical composition is:

Wheat flour	30 per cent
Bengal gram flour	30 per cent
Edible ground nut flour (defatted).	10 per cent
Jaggery	30 per cent
Vitamin, and minerals	Fractional quantity.

Process of manufacture involves pre-cleaning of raw materials, roasting under appropriate conditions, powdering to required mesh size and mixing the ingredients in required proportion and packing in air tight PVC bags. Equipment like destoner, roaster, pulveriser, mixer and heat sealer are easily available.

Economics

Capacity of the unit	300 kg/day
Building	100 Sq.m.
Equipment	Rs. 3.00 lakh
Working capital	Rs. 2.00 lakh
Profitable	30 per cent

Utilities

Power requirement	10 KW/day
Water	200 kl/day

Iodised Salt

Salt is essential for human beings as well as animals. Iodine deficiency in food gives rise to diseases like goiter and birth of mentally retarded children. Iodisation of salt is necessary to save population from these diseases. Multinational and big companies are manufacturing iodised salt and marketing it at high price. It is possible to produce iodised salt at cottage scale level for marketing at affordable price.

Although various processes are available but immersion process is easy to operate. Thirty gm of potassium iodate ($KI0_3$) is dissolved in 200 l of water. Raw salt is immersed in the water for 5 minutes and then transferred to concrete platform for drying to drain out excess water. It is kept overnight for slow drying. Subsequently, it is ground in roller grinder and then packed in polythene bags. This water-iodine solution is sufficient for 1000 kg of salt to attain 30 ppm of iodine in salt. Some iodine is lost during processing. However, desired level of 15 ppm should be there during the shelf-life of salt.

Techno-Economics

Capacity–1T/day–300 working days.

1. Land	rental
2. Machinery/equipment (Roller grinder, concrete tank and bamboo baskets)	50,000
3. Working capital (two month salary, Raw material)	1,00,000
	1,50,000

Sales (300 T × Rs. 3/kg) =	Rs. 9,00,000
Less:	– 6,00,000
	3,00,000

Test kits are available in the market to test the iodine content in the salt.

Drying and Powdering of Red Chillies

Traditionally the fresh red chillies are sun dried (with initial moisture level of 74-75 per cent) to a moisture level of 15-16 per cent to enable their grinding into powder. But this process is very time consuming as it takes 15-20 days to dry these to the desired moisture level.

CFTRI, Mysore has developed a faster drying process called "Dipsol process" which involves soaking fresh chillies in an emulsion of anti-oxidants for 5 minutes, draining the emulsion and drying the chillies in multi-tier wire net trays at the rate of 5-10 kg per square meter of the tray at ambient temperature. The treated material thus dries in a week. However, if solar drier is used it takes only 3 days. It is found that 7.5 L of disposal emulsion comprising (hydroxy bet anisol, potassium carbonate, coconut oil and water) are required to treat 100 kg of chillies and approx cost is Rs.30.00 *i.e.* 30 NP/Kg.

Salient Features

1. Faster drying
2. Requires less space
3. Better retention of colour and pungency
4. Gives hygienic super quality product
5. Gives high yield of finished product (2 per cent more) due to minimum breakage and loss of seeds.
6. Additional cost of treatment is nominal.

Dehydration of Fruits and Vegetables

The drying of perishable fruits and vegetables in order to preserve these during season of abundance and making these available during lean or off season is the most economic and safe method of preservation for use in home, large scale catering establishments, defence rations, hotels restaurants etc. Dehydration implies removal of moisture under controlled temperature and humidity conditions, without impairing the nutrient content and palatability. Removal of water or moisture content to desired level

Drying Schedule for Some Vegetables

Vegetable	*Preparation*	*Pre-treatment*	*Max. Drying Temp (°C) in Home Drier*	*Drying Equipment*
Potatoes	Peel and cut into 1 cm thick slices	Blanch in boiling water or steam for 3-4 mins and in sulphite in 0.12 per cent potassium meta-bi-sulphite (KMS) solution for 10 mins using 0.5 kg of solution/kg of slices	65	Solar or Hot air
Carrots	–do–	Blanch in boiling water or steam for 3-4 mins and in sulphite in 0.25 per cent KMS solution for 100 mins using 0.5 kg of solution per kg of slices	60	– do–
Cabbage	Remove outer leaves and cores and cut into 4-8 mm thick shreads	Blanch in boiling water as above for 5-6 mins and in sulphite as above.	55	– do–
Cauliflower	Remove stalks, covering leaves and stems, break apart flowers and cut into 10-12 mm thick pieces	Blanch as above and in sulphite in 0.125 per cent KMS solution using 0.5kg of solution per kg of material.	50	– do–
Onions	Remove tops and tails peel and cut into 4-8 thick shreds	Nil	55	– do–

Contd...

Table Contd...

Vegetable	*Preparation*	*Pre-treatment*	*Max. Drying Temp (°C) in Home Drier*	*Drying Equipment*
Garlic	Peel the cloves and cut into 6 mm thick shreds	Nil	55	Vac. drier
Okra (Sliced)	Remove both ends	Blanch in boiling water for 4-5 mins and in sulphite in 0.25 per cent KMS solution for 10 mins using 0.5 kg of solution per kg of slices	60	Hot air drier
Okra (Small sized 50-75 mm length, whole)		Blanch in boiling wter for 6-7 mins	60	– do–
Bittergourd	Remove both the ends with stainless steel knife and cut into 6mm thick slices	Blanch in boiling water for 8 mins	60	– do–
Pumpkin			65	– do–
Beetroot			60	– do–
Green peas	Remove the pod and the prick with a small nail or pin	Blanch for 3-4 mins in boiling water containing 0.5 per cent KMS 0.1 per cent magnesium oxide and 0.1 per cent sodium bicarbonate.	60	– do–

Contd...

Table Contd...

Vegetable	*Preparation*	*Pre-treatment*	*Max. Drying Temp (°C) in Home Drier*	*Drying Equipment*
Split field beans	De-skin and split whole field bean (Dolichos) seeds by boiling in 1 per cent sodium carbonate solution for 5-7 mins and rubbing off the skin under spray of water	Blanch as above for 2 mins.	60	Hot air drier
Spinach	Sort, wash thoroughly and cut into 10-12 mm portions using a stainless steel knife	-do-	60	– do–
Fenugreek	Sort, wash thoroughly and remove stalks and stems	-do-	60	– do–
Banana	Wash, peel cut into 12mm thick slices	Nil	40-48	– do–

prevents spoilage of fruits and vegetables. Common items which are dehydrated are potato, green peas, carrot, cauliflower, cabbage, green fenugreek and spinach, grapes, Banana, Figs, apple, unripe mango, ginger, garlic, onion, okra etc.

Various methods of drying are used such as hot air driers, vaccum shelf driers, freeze driers, spray driers, drum driers, osmotic dehydrator, solar drier etc. Solar drier is the cheapest and a energy saving device which can be easily adopted for most of the vegetables. Vaccum shelf drier is used to preserve aroma for products like garlic.

Techno-economics

Building capacity 50 sq.m.	1.00
Equipment/machines	2.00
Working capital	1.00
Project cost	4.00
Profitability	40 per cent

Sunflower Seeds

Sun flower seeds are used for producing edible oil rich in unsaturated fatty acid, snacks, chikkis, confectionery items, edible quality cake/flower etc. Present production of seeds is about 1.0 million tonnes and is expected to increase in view of nutritional importance of sun flower oil. Sun flower is mostly cultivated in states of Karnataka, Tamil Nadu, Andhra Pradesh, Haryana, Punjab, Maharashtra and U.P.

Proces Flow Chart

Sun flower seeds ⟶ Cleaning ⟶ Grading ⟶
Decortication ⟶ Kernels ⟶ Air separation

Economics

Capacity	500 kg/day
Building	50 sqm.
Equipment	4.00 lakhs.
Working capacity	2.00

Yield is 62-63 per cent kernels from seeds.

Utilities Per Day

Manpower	5 Nos.
Power	4 KW
Water	10,000 L.

Potato Chips

Potato chips or wafers are one of the most popular items consumed in almost all parts of India. Currently potato chips are being manufactured by multi-nationals and large firms that have captured sizable market earning heavy profit margin. The cottage/ home scale units in semi urban and rural areas also have their market share. Cottage scale units can enlarge their market by improvement of quality through little mechanisation, adoption of better packaging practices and by offering cheaper sale price.

The process of manufacturing involves washing, peeling, slicing in the thickness range of 1/8" to 1/16". After peeling, wash the slices to de-gelatinize and avoid sticking. Sliced potatoes are blanched for 3 to 4 minutes in boiling water and then soaked in 0.125 per cent potassium metabisulphate solution for 10 minutes. Subsequently, the slices are dried in sun or by using solar drier/hot air drier. The drying period is 6-8 hr. Moisture content of the dried chips/strips should not exceed 7 per cent. The dried chips could be packed for marketing or deep fried and mixed with salt/spices/ mint powder etc.

Capacity	50 kg/day of chips
Building	50 sq.m.
Equipment	Rs.1.50 lakh

Peeler, slicer, brancher of SS or
Al, deep fat frier, basket
sieve, pvc sealing machine)

Working capital	Rs. 1.00 lakh
Power	4 KW/day
Water	2000 L/day
Fuel diesel	20 L/day.

Papad

Papad is an improtant part of food as well as a popular snack. It has demand in rural as well as urban areas. In the traditional method papad is prepared from dall flour and spices by rolling with a rolling pin which requires lot of energy. In Gujarat some power operated rollers are in use. CFTRI Mysore has developed foot as well as hand operated paper press which minimizes the effort and enhances the production capacity.

The thickness of the papad can be adjusted by adjusting clearance between the pressing dices. The quality of the papad made with these presses is no way different from the traditionally hand-rolled papads. This machine is also suitable for making papad from horse gram which is hard and difficult to roll by roller pin.

Features

Hand operated method	60 papads/hour.
Hand operated fly press	200 papads/hour.
Pedal operated press	500 papads/hour

Simple Wheat Milling Process for Bakery Flour (Maida) and Atta

Presently about 80 per cent of the total annual production of wheat is processed in large roller flour mills in organised sector and 20 per cent is processed by disc mills (chakkis) in urban and rural areas. While roller flour mills enable one to produce atta, maida and suji/rawa the chakkis produce atta only. Recently, trends have changed and large population is switching over to packed easy to use Atta. CFTRI, Mysore has developed a mini flour mill which enables one to get all three products–maida, suji and atta. Many multinational companies and large business houses are entering the food industry. It is felt that if substantial portion of agro-processing is carried out in rural areas there is possibility to generate employment for millions and improve the economic standard of rural population.

CFTRI, Mysore has developed number of mini agro processing machines which could be set up in rural areas for this purpose. One such initiative is to set up mini flour mill for employment generation and value addition of agricultural produce. In this machine, all the steps used in roller flour mill are performed manually such as–

- ✰ Cleaning of wheat by using sifters/graders.
- ✰ Conditioning, tempering of wheat by moisturing the wheat and keeping for 20 hours.
- ✰ Bran removal by Hullers
- ✰ Grinding of de-branned grain in stainless steel aberrated steel disc mills.
- ✰ Sieving–Separation of atta and maida by using gravity sifters.

All the machines like clearner, huller, grinders and sifters are available in the country from licensed manufacturers–.

Economics

Capacity	250 kg/hour or 2T/day in a single shift.
Machinery	Rs. 3.50 lakhs.
Shed	1000 Sq.ft.

Yield

Bakery flour	350 kg.
Chapatti atta	560 kg.
Bran	60 kg.
Milling losses	20 kg.
Refractions	10 kg.
	1000 kg.
Connecting load	20 kw, 3 phase.

Advantages

- ✰ Low capital cost
- ✰ Potential for self employment for others.
- ✰ No need of skilled labour.
- ✰ All the nutrients are preserved.
- ✰ Custom milling

- ☆ Bran could be used for cattle feed formulation.
- ☆ Adoptability of the process for suji milling of durhum wheat.
- ☆ Avoids use of emery discs.

Machinery Manufacturers

1. Mysore Industries, Mysore.
2. Nalanda Agro Works, Industrial area, Nalanda Nagar, Khurji, Patna, Bihar.

Mini Rice Mill

India is the second largest producer of rice in the world. The mechanised sector of rice milling industry handles more than 45 million tonnes of paddy annually. About 10-15 million tonnes are processed through large scale modern mills involved in producing fine quality of rice for export market. The major portion of paddy is still processed through hullers which are usually low capacity mills and result in very high percentage of brokens. In these hullers both shelling and polishing operations are carried out simultaneously and there is no control on the polishing of rice. As a result impure bran mixed with husk is obtained and a higher breakage of rice results in loss of revenue. To overcome these problems it is necessary to carry out the shelling and polishing in two separate units. A mini rice mill has been developed by CFTRI, Mysore and many manufacturers have sprung up having their own versions. The sheller is a compact unit designed on the densimetric classification, principle; the polisher could be light either vertical cone polisher or a horizontal roller polisher.

The salient features of mini rice mill are:

- ☆ 1-4 per cent extra yield of head rice depending on the variety
- ☆ Production of pure rice bran free from husk.
- ☆ Control over degree of polishing from 3 per cent onwards.
- ☆ Capacity as low as 500kg of paddy/hour and can also serve as custom milling unit.

The pure bran can be supplied to solvent extraction industry after stabilisation for production of edible rice bran oil. Chemical stabilisation process is easy to operate compared to steam or roasting treatment. The principle of the process lies in maintaining pH of the bran to the level where the lipase activity would be negligible and within safe limit for FFA which is 2-4 per cent for edible oils. Commonly available hydrochloric acid is sprayed over bran which inactivates enzymes to nearly zero level.

About 40 litres of commercially available concentrated hydrochloric acid is required for one tonne of bran. Mixers of drum type are used for spraying HCL over bran.

Equipment/machinery is available from the following manufacturers:

- ✰ Mysore Precision Engineers, C-123-124, Industrial Estate, Yadavgiri, Mysore.
- ✰ Kisan Krishi Yantra Udyog, Kanpur, U.P.
- ✰ Vishwakarma Iron Foundry, Kashmiriya, Tanda, Dist. Ambedkar Nagar, U.P.
- ✰ Super Engineering Works, Howra-Amta Road, Dass Nagar, Howrah.

Economics

1. Land 500 Sqm.	Rs. 50,000.00
2. Work shed 100 Sqm.	Rs. 2,00,000.00
3. Machinery	Rs. 1,00,000.00
	Rs. 3,50,000.00

Mini Dall Mill

CFTRI, Mysore has developed mini dall mill to dehull pulses as an appropriate technology for rural sector. At present dall mills are set up in semi-urban and urban areas involving heavy investment. The pulse seed from rural areas is brought to urban areas and again substantial quantity is transported back to meet rural needs of dall as well as husk, powder and bran for use as cattle feed in rural areas.

The machine comprises a grader, emery coated cone, pneumatic exhaust for collecting husk and powder and an aspirator for separation of dall, brokens etc. It provides employment to 3-4 persons per unit which will help in solving unemployment problem.

After cleaning the pulse, seed is soaked in water for the duration depending upon variety of seed. Subsequently, it is dried to moisture level of 11-12 per cent followed by grading in sieves of 3,4,5mm screens. The graded material is fed to conical drum which rotates at 90-100 rpm. Due to abrasive action between cone and wire mesh the pulse seed is dehusked and split followed by separation in aspirator. The pulse is then treated with 2 per cent edible oil for 3-4 hr and then sun dried for 3-4 days before packaging for market.

Special Features

Capacity of the mill	100-150 Kg per hr
Space required	2 m x 2 m.
Yield of dall	77-80 per cent
Dehulling efficiency	96-99 per cent
Brokens	1-2 per cent
Power requirement	1 h.p. single phase.

Economics

Land and Building	Rental
Machine cost	Rs. 1.00 lakh
Working capital 15 days	1.00 lakh
	Rs. 2.00 lakh.

Profit–20 per cent on investment.

Machinery Manufacturers

- ☆ AMI Engineering, Station Road, Opp. Veena Cinema, Patna-800 001 Bihar.
- ☆ Nalanda Agro Works, Nalanda Nagar, Khurji, Patna–800 010.
- ☆ Vishwakarma Foundry Works, Kashmiriya, Tanda–224190, Ambedkar Nagar, U.P.
- ☆ Murugan Industries, St. Site No.68/W, Hootagali Industrial Area, Belavadi Post, Mysore-571 186.

Modified Oil Expeller

Traditional edible oil industry makes use of Kohlus for extraction of oil from oil seeds. The nutritional quality of oil is better but yield is low. At the same time introduction of mechanical expellers increases the oil yield but nutrients are burnt due to expelling of oil at temperatures as high as 80-90°C. Especially in the case of mustard oil, pungency is lost due to evaporating of allye iso-Thiocynate (essential oil). In view of the merits and demerits of expellers, Mechanical Engineering Research and Development Organisation (MERADO) Ludhiana has developed improved oil expellers where the temperature of expelling zone is kept below 60°C by providing water jacket around the expelling zone. Stainless parts were incorporated to reduce maintenance cost. Thus it is possible to get oil of high pungency and higher yield.

Oil expellers of ITPD and 6TPD are now available from licencees of MERADO. This expeller is suitable for oil seeds like mustard, ground-nut, linseed, niger, sunflower and even neem seed. The cost of ITPD expeller is Rs. 1.15 lakh and it is suitable for small entrepreneurs.

Special Features

- ☆ Extraction of pungent oil from mustard seed due to extraction at low temperature.
- ☆ High extraction efficiency.
- ☆ Automatic cake conveying system.
- ☆ Longer life of critical components.

Capacity	1 ton/24 hr. (mustard)
Power required	7.5 H.P.

Plant Suppliers

- ☆ Tegh Engineering Works G.T. Road, Miller Gunj, Ludhiana.
- ☆ Swastic Traders G.T. Road, Miller Gunj, Ludhiana.
- ☆ Jash Agro-tech, Miller Gunj, Ludhiana.

Mushroom Cultivation

Till 1980 mushroom cultivation was restricted to hilly areas of Northern states. Subsequently the plains of the North started growing the crop on commercial scale. Among different states in India, Himachal Pradesh contributes around 2500T, Haryana 1650T, Punjab 1200T and Uttar Pradesh 2500T. Including Jammu and Kashmir, Karnataka, M.P., Maharashtra etc. the estimated production is 25000T. Indian entrepreneurs have now realized the nutritive and medicinal value of mushroom and many units have come up having controlled conditions of humidity and temperature.

The World consumption of mushroom has grown 10 times during the last 30 years. At present the global demand is 27 lakh tonnes while production is 20 lakh tonnes. As such there is potential for export to Switzerland, France, Germany etc. India has all the requisites to become a leading mushroom producer due to availability of low cost labour, suitable temperature, humidity and plenty of agricultural residues. Mushrooms are used for their medicinal value apart from use as nutritive food, slimming, agent and in aneamia, etc. Over 2000 species of mushroom have been identified in the World but only 20 species have been selected for commercial cultivation. Commonly cultivable species are:

- ☆ Pleurotus (Oyster mushroom)
- ☆ Volvariella (Paddy Straw mushroom)
- ☆ *Agaricus bisporus* (Button mushroom)
- ☆ Shitake etc.

Mushroom is a simple form of life known as fungus. It lacks chlorophyll and can not, therefore, make its own food. It, therefore, grows on dead organic matter or parasitically or symbiotically with other organisms. The edible part constitutes the fruit body. It varies widely in size and shape in different mushrooms. These are mostly produced above the substrate from which the food material is absorbed through the network of white threadlike structures.

Mushroom has high nutritive value, which has been recognized in recent times although its use as a delicacy for flavouring purpose has been known since ancient times. It is a good source of high quality protein, vitamins and minerals which is higher than those in cereals, fruits and vegetables. In addition to protein content of 40 per cent they contain carbohydrates-2.4 per cent, fat-0.4 per cent minerals-0.6 per cent and water-91 per cent on fresh weight basis. They also contain vitamin B, C, D and K and minerals like calcium, phosphorus, potassium, iron and copper. Mushroom is devoid of starch and hence forms a good diet for diabetic persons. Its low fat content, rich in linoleic acid free from cholesterol makes a good diet for blood pressure and heart patients. Mushrooms, due to alkaline ash (high sodium and potassium) and high fibre content are highly suited to the patients suffering from of hyperacidity and constipation. Regular intake of mushroom helps to reduce obesity.

Cultivation of Oyster Mushroom (*Pleuotus sajor-caju*)

Oyster mushroom is a cellulose loving fungus and can be grown on variety of celluloytic materials including paddy straw, wheat straw, bajra stalks, maize stalks, maize cones and number of other agricultural waste materials. Different species need different temperature conditions for their growth and hence they are ideally suited for cultivation at different times of the year in various regions. The temperature requirement is higher (20-30°C) and relative humidity of 80-90 per cent is required and hence this mushroom possesses good potential in our country, especially as the technology for its cultivation recently perfected is very simple and cheap. Further, the produce has a longer shelf life and can be dried easily. The technique for its cultivation involves the following steps.

Preparation of Beds

This mushroom proliferates equally well on both wheat and paddy straw. Selection of straw is made depending upon availability of straw as substrate which is easily available in rural areas. It should be preferably fresh or not more than a year old and should not have been exposed in rains after cutting. For the preparation of substrate, and following items are required.

1. An empty drum or tub
2. Formalin
3. Bavistin
4. Paddy/Wheat straw
5. Polythene bags (24" x 18")
6. Water

One hundred litres of water (approx. 10 buckets) is taken in the drum then 100ml. of formalin and 7 gm. of bavistin are added in it. It is properly mixed with a stick. Ten kg. of dry straw is then wetted in this chemical solution for 24 hr. Next day straw is taken out and spread on a clean floor. It is kept as such for up to one hr so that excess water drains off. The weight of wet straw will now be approximately 42 kg.

Spawning

Chemically treated wet straw is thoroughly spawned @ 10 per cent on wet weight basis of the straw. After spawning this material is transferred to perforated polythene bags having 8-10 holes with 1cm dia. Mouth of the poly bag is closed by a rubber band. They are now arranged on raised platform keeping a distance of 15-20 cm from one another.

Removal of Bags for Production of Mushrooms

After about 20 days the compost is fully covered by the fungus. The polythene bags are removed by turning them side up down and tapping the blocks on the sides. These are again arranged on a wooden plat form. Watering is done twice a day or as required to ensure 80-90 per cent RH. Walls and floor are also sprinkled with water. Aeration has to be carried out for 1-2 hr daily. Sufficient amount of diffused light is required for normal fruit body formation. Mushrooms will appear within 4-6 days which can be harvested after 2-3 days. The right stage of picking is up curving of the margins of the pileus or before shedding of white spore powder. Flush will appear within a period of 4-5 weeks.

Harvesting and Marketing

The fruit bodies arise in clumps and form tiers on the substrate. They are variable in size, shape and color. They should be harvested carefully to avoid any disturbance to other pinheads as much as possible. These are then packed in perforated polythene bags for sale in fresh form. Extra produce can be sun dried or dehydrated at 50-55° C and stored in sealed polythene bags. It is also used for making pickles and soup. Solar driers can be used for dehydration of oyster mushroom.

Economics

Cost of production of this mushroom will be about Rs.24-25 per kg and from one kg dry straw about 600 gm of fresh mushroom can be harvested. In another method of cultivation wetted wheat and paddy straw or any other above mentioned substrate is boiled in hot water (80°C) for 1-2 hr. After it is cooled down it is spawned @ 2-3 per cent. Other steps involved are the same as mentioned for the above technique.

Capacity	7-8 kg/day or 1560kg/year in 8 months
Capital investment	Rs.5000/-
Working Capital	17,000/-
Cost of production	Rs.24-25/kg
Selling price	Rs. 35/- kg.
Profit	Rs.10 per kg.

Process Flow-Sheet

Preparation of compost/Substrate

↓

Spawning

↓

Filling in polythene bags

↓

Cropping

↓

Harvesting

↓

Packaging in polybags-fresh → Dehydration

↓ (Packaging in polybags-fresh) ↓ (Dehydration)

Marketing ← (from Dehydration)

Source of Technology: National Mushroom Research Centre (ICAR) Solan, Himachal Pradesh.

Agricus bisporus (Button Mushroom) cultivation requires (i) preparation of compost on concrete platform which takes 28 days, (ii) transfer of compost to wooden trays (3'x2'), (iii) spawing of compost (iv) casing of trays with light soil, (v) Frequent watering using sprinklers. Sprouting takes place after 4 weeks and continues for a month. From one tray of 3'x2' size one can get yield of 3-4 kg of Button Mushroom. The crop is harvested with sterilised scissors or knives and packed in polythene bags. The fresh button mushroom has a shelf life of 3 days. Canning in brine water enhances its shelf life upto 18 months.

Button Mushroom requires very hygienic space having environmental conditions of 11-18° C temperature and 85-90 per cent humidity. Normally four tier system of racks is used to economise space, 400 trays of 3'x2' size are kept in room space of 1000 Sq.ft.

Papain from Papaya

Papaya is a versatile plant having number of uses and enzymatic properties. It could also be called medical plant rendering high income to farmers. One can manufacture products like tutti-frutti, candy slices (murabba) and dehydrated papaya powder from raw papaya. People suffering from ulcers and other stomachic problems eat ripe papaya. A longitudinal cut is given to unripe papaya fruit in the evening and it is covered with polythene. Next day, due to heat of rising sun, the latex (milk) starts oozing out from cut portion. It is collected in the polythene bags, which is then washed, dried in vacuum shelf driers and finally powdered for formulation of papain of IP or B.P. grade. The papain is used in applications like tendering of meat for fast cooking, softening of silk fibre and leather, chill proofing of beer and pharmaceutical preparations for curing stomachic ailments. Papaya pulp is also used in food processing industry.

Farmers generally cultivate CO, CO_2 developed by TN Agriculture University and Taiwanese varieties. One plant gives an yield of about 150 kg of fruit in 8-10 months period. Plant to plant distance of 2 meters is kept. Some farmers adopt inter-cropping practice and grow chillies for higher return. One can generate a profit of 10-12 lakhs from a plantation of 20 acres. A small plant for production of papain could be set up near the field which costs around Rs.5.00 lakhs on machinery apart from infrastructure like shed, utilities like water and electricity. Technology for papain production could be obtained from CFTRI, Mysore.

Shed 2000 Sq.ft.	3.00 lakhs
Plant (vacuum shelf drier, grinder packaging machine	5.00 lakhs
Working caital	1.00 lakhs
	Rs. 9.00 lakhs

Sugar Cane Juice

Sugar cane juice is a very popular drink in India and other Asian countries. At present it is being produced by vendors using hand operated or motorised crushers and served within a short time. The juice can not be stored even for a couple of hr due to initiation of fermentation process. The juice is produced and served under very unhygienic conditions and hygiene conscious public does not opt for sugar cane juice. Hygienically produced and bottled sugar cane juice could find use as other soft drinks and has good potential market as the juice is nutritious containing minerals like iron, magnesium, phosphorous and calcium and organic acids like malic, succinic and acetonic. In Ayurvedic system of medicine, sugar cane juice is recommended for use by jaundice patients.

In the process of manufacture first step is to clean sugar cane by soaking in water for an hour, scrubbing with nylon-coir brushes for cleaning the sugar cane and then dipping in 1 per cent potassium metaby sulphite solution for 5 minutes. The cleared canes are then crushed in pre-cleaned crushers, filtered through filter cloth and collected in SS containers (rinsed with 1 per cent K.M.S.) Preservatives are added to the juice which is finally filtered through filter press. The juice is diluted to 15 Brix by using deionised water. The juice is then blended with flavours like lime-lemon, ginger capsicum etc. The blended juice is filled in 200 ml bottles, crowned and then pasteurised at 70°C for 10 minutes. The bottles are now cooled and stored for supply.

Investment of Rs. 16.00 lakhs is required for a capacity of 6000 bottles per day. Technology is available from CFTRI, Mysore or Agro Processing Division, Tamil Nadu Agriculture University, Coimbatore.

Chapter 4

Medicinal and Aromatic Plants

The use of medicinal plants for health care is as old as human civilization. The oldest repository of human knowledge is 'Rigveda' written in era 4500-1600 BC. Later 'Atharvaveda' mentions about 8000 formulations used in 'Ayurveda'. CSIR has digitalised this knowledge base for wider applications throughout the world. In the 20th century there was perceptible change from herbal medicines to synthetics due to R and D in western world but herbal medicines continued to cater to the needs of Third World countries. There has been renewed interest in herbal drugs due to their unique properties and cheaper cost and thus bringing them within the reach of common man and Organisations like WHO have recognised the potential of folklore/ethno system of health care and included herbal drugs in "National Health care programmes". In global context it is necessary to define the term medicinal plant. In broader sense all herbal plants have been credited with having medicinal properties–effects that relate to health or which have proven to be useful as drugs by western standards or which contain constituents that are used as drugs. It is estimated that there is Rs. 40,000 crores world market for medicinal and aromatic plants which is likely to grow due to low toxic effects of medicinal plants. There is export of above Rs.4000 crores of medicinal plant from India.

A large number of medicinal and aromatic plant raw-materials have been collected from the forest sources but due to rise in demand cultivation has been taken up in various parts of the country. Earlier traditional suppliers and traders dealt with the industry. Now, CIMAP and others have published information for the benefit of growers who can now directly supply plant material to pharmaceutical industry. National Medicinal Plant Board (NMPB), New Delhi and National Horticulture Board, NABARD and other nationalised banks provide financial assistance for cultivation of medicinal plants.

In alternate system of medicine very large variety of plant materials is used for ayurvedic formulations. Following major medicinal plants are required for commercial supply to pharma companies.

Sl.No.	*Name of Plant*	*Drug*	*Use*
1.	*Acorus calamus* (Vacha)		Tranquilizer
2.	*Adhotoda vasica* (Vasa)		Oxytocic
3.	*Andrographis paniculata* (Kalmegh)		Bacillary dysentry
4.	*Asparagus racemosus* satavar		Anti-oxytocic, anti-ulcer, lurv and treatments of burns wound healing enhance lactation.
5.	Belladonna		In peptic ulcers, intestinal disorders, colic and urinary bladder pains, kidney stones.
6.	Cinchona		Quinine–anti malarial property
7.	Datura		Source of tropane alkaloids.
8.	Digitalis		Digitoxin–congestive heart failure, digitalis–cardiac muscles.
9.	*Discoria composita floribunda*		Diosgenin precursors for synthesis of cortisones, sex hormones and oral contreceptives
10.	Ergot (Ergot of rye)–(fungus on rye plant)		Alkaloids, analgesic for migrane (ii) Hypertension and other peripheral circular disorders.

Contd...

Sl.No.	Name of Plant	Drug	Use
11.	*Hyoscyamus* (Henbane)		Relieve spasm of urinary tract with strong purgative to prevent gupping.
12.	Liquorice		Flavouring agent, demulcent and mild expectorant, anti-spasmlytic and anti ulcergenic.
13.	Opium poppy	-	Morphine–relieves from extreme pain-codeine-analgesic papa-verine–smooth muscle relaxant in asthmatic, gastric and intestinal spasms and coronary ailments.
14.	Periwinkle (*Catharanthus roseus*)	-	Anti-cancer drugs–vincristine–vinblastine.
15.	Psyllium seed	-	Chronic constipation, also controls diarrhoea and dysentery. Seed mucilage used as stabiliser in icecreams, chocolates and other food materials.
16.	*Artemisia annua*	-	Two derivatives Artemisinin and arteether-very effective against chloroquine resistant strains of malaria parasite as well as against cerebral malaria.
17.	*Rauvolfia serpentina*	-	Sedative–alkaloides–reserprine and vescinnanins are used in modern medicine as tranquiliser and hypertensive.
18.	*Senna- Casia Augustifolia*	-	Laxative
19.	(i) *Bacopa monnieri* (Brahmi)		Antianxiety agent improves intellect and memory.
	(ii) *Centella asiatica*	-	Vulnerary wound healing.
20.	*Casia papaya* (Papita)	-	Proteolytic mucolytic.
21.	*Curcuma longa* (Haldi)	-	Choleretic and other usages
22.	*Picrorhiza kurroa* (Kutki)	-	Anti-hepatotoxic.
23.	*Cosciniun Fenestralim* (Dam Haldi)	-	Cholera and gastroenteritis.
24.	*Taxus baccata*	-	Anti-cancer drug (Taxol)
25.	*Glorisoa superba*	-	Colchicine
26.	*Mappia foetida* (Ghaneru)	-	Anti-cancer

Contd...

Sl.No.	Name of Plant	Drug	Use
27.	*Withania Somnifera* (Ashwa gandha)	-	Tonic for vitality
28.	*Aloe vera*	-	Anti-septic, cosmetics and many medicinal uses.
29.	*Stevia*	-	Sugar free, control of blood pressure.
30.	*Glycemia- gudmar*	-	Anti-diabetic
31.	*Acolius, Pathar chur*	-	Urinary and Kidney stone
32.	Blue green algae	-	Vitamin A,B,C,E and Biotins
33.	Guggul (*Commiphora wightii*)	-	Lowering of cholestrol, anti-arthritis.

Aromatic Plants

Mentha

Soil of Climate	*Planting Material Season*	*Irrigation*	*Inter Crops*
Deep loam pH 6.0-7.50	350-500 kg/ha Jan/Feb.	Frequent, avoid water logging	Maize, paddy, potato, Arhar

Sl. No.	*Mints (Varieties)*	*Cost of Cultivation Rs/ha*	*Oil Yield kg/ha*	*Price Rs/l*	*Cost of Oil Produced*	*Net Return*
1.	Mentha arvenis	18000	150	350	52500	34,500
2.	Pepper mint	18000	100	400	40000	22,000
3.	Bergamot mint	18000	150	300	45000	27,000
4.	Spear mint	18000	125	300	37500	19,500

Uses

Used in production of menthol, confectionary and perfumery products.

Cymbopogons

	Soil and Climate	*Planting Material and Season*	*Irrigation*	*Yield of Oil*	*Profit Rs.*
Java citronella	Sandy loam to loam pH 6.0-7.50 Temp. 9-35ºC	5-7 cm long slips July-Aug/ Feb-March 55000 slip/ha.	20-25 irriga-tions in a year	150-200 kg oil/ha from 4 cuttings. Rs. 350/kg	40,000/ha.
Lemon grass	Rich loam to poor laterite sandy loam with rich manu-ring fine tilth.	Slips- 62500/ha CKP-25 variety	Frequent irrigation	200-250 kg/oil from 4 cuttings Rs. 400/kg	50,000/ha.
Palma-rosa (martini)	Sandy loam to rich loam 30-40º C rH 78%	Seeds and slips– 60,000 Jan/Feb July/Sept.	20-25 irrigations	125 kg from 4 cuttings Rs. 600/kg	35,000/ha

Uses

Soaps, cosmetics, fragrance and flavours, B-ionone from lemon grass. *Java citronella* is also used as mosquito repellant.

Vetiver

	Soil and Climate	*Planting Material and Season*	*Irrigation*	*Yield of Oil*	*Profit Rs.*
Vetiver (khus)	Loamy-cum sandy soil (pH 9.5-10.0) warm and damp climate slopy and undulated soils for soil conservation (KS-1 variety)	60,000 plants/ha	Frequent irrigation	15 kg of oil @ Rs.15000	1,50,000/ha in two years

Used

Food flavours, soap, perfumery and chewing tobacco industries.

Geranium

High Grade perfumery product having sweet rosy odour now being cultivated in Nilgiri and Palni Hills in South and Uttarakhand.

Soil/Climate	*Planting Material + Season*	*Irrigation*	*Yield*	*Profit*
Suitable for Hilly areas, well drained sloppy soil pH 6.7-7.5 temp of 25-30ºC	Green cuttings having 3-4 nods Nov-Dec (in south) Feb-March (north) 40,000 plants/ha 10x25 cm space	Light irrigation	25-30 T herb or 45-50 kg/oil/ha	Rs 70,000/yr

Tagetes minuta (Marigold)

Soil/Climate	*Planting Material + Season*	*Irrigation*	*Yield*	*Profit*
Sandy loam to clay loam pH 4.5-7.5 temp. to semi temp. climate	2 kg seed/ha in October 40-45 cm plant to plant	6-7 irriga- tions in planes 3-4 irri. in Hilly areas	300-350 q/ha 65kg oil @ 1500/- per kg.	85000/ha deducting cost of cultivation

Uses

Medicines, aroma chemicals and pesticides.

German Chamomile

Soil/Climate	*Planting Material + Season*	*Irrigation*	*Yield*	*Profit*
Moist, mod heavy and fertile soils, calcarous soils pH 8 to 9, winter crop in N. India ht. 100-1500 m. temp. 18ºC-20ºC most suitable	500-800 g seeds/ha. Oct-Nov. in plains, May- June in Hilly areas.	Frequent irrigation	10q. flowers @ Rs. 50/- kg.	25000/ha.

Uses

Herbal tea, toothpaste, mouth wash, lipstics, medicated soaps, skin and sun burns, baby oils and shampoos.

Rose Oil

Soil/Climate	*Planting Material + Season*	*Irrigation*	*Yield*	*Profit*
Sandy loam to clay loam, pH 6.0-7.5, well drained rich in organic matter.	10,000 rooted cutting/ha 1.0-1.15m × 2.5-3.0m. Oct-Nov.	Frequent irrigation	2-3 ton petals 0.03% oil	Rs.1,00,000/ ha.

Use

Perfumery industry:

Patchauli

Soil/Climate	*Planting Material + Season*	*Irrigation*	*Yield*	*Profit*
Partially shaded well drained fertile soil. Humid and warm climate.	Through cuttings, having 4-6 leaves	Distributed	8000 kg leaves 1600 kg dry leaves- 40kg oil.	Rs.30,000/ ha.

Use

Perfumery industry as a fixative.

Ocimums–Four Varieties

Soil/Climate	*Planting Material + Season*	*Irrigation*	*Yield*	*Profit*
Moderate fertile well drained soil Tropical climate	Seedlings from Nursery in July	5-6 irrigations after rains.	85 kg oil/ha.	Rs12,000/ha

Uses

Condiment, cosmetics, perfumery, medicines etc.

Source of Technology

1. Central Institute of Medicinal and Aromatic Plants (CIMAP) P.O. CIMAP, Lucknow.
2. Regional Research Lab, Canal Road, Jammu Tawi.
3. Regional Research Lab, Bhubneswar.
4. Regional Research Lab. Airport Road, Jorhat.
5. Institute of Himalayan Biosphere Technology, Palampur, H.P.

There are 10,000 pharmaceutical units in the country to whom growers can directly contact for supply of medicinal herbs apart from traders and exporters who deal with marketing of medicinal plants.

Manufacturing Process of Juice, Gel and Powder

Aloe vera

Aloe vera has its origin in the African continent. Its thick leaves contain the water for supply to plant to survive long periods of drought. The plant can survive very harsh drought where most other vegetation disappears. The first written reports on *Aloe vera* are more than 2000 years old. Alexander the Great had special interest in parts of Africa where mainly aloes grew. He used them heal wounds of his soldiers. Cleopatra used it as a skin care product and bible mentions the use of Aloe.

During the nineteenth century *Aloe Vera* plant was brought from Africa to other tropical climates for cultivation such as Caribbean islands, India, Venezuela and Mexico.

Aloe vera Gel contains 99.3 per cent water. The remaining 0.7 per cent are solids that consists of polysaccharides of the glucose and mannose type. Together with the enzymes and amino acids in the gel they give the gel special properties of skin care. The gel stimulates cell growth and restoration of damaged skin. It moisturises the skin because of its water holding capacity and cooling effect. *Aloe vera* gel is now widely used in cosmetics all over the world. It has been used for treatment of skin lesions caused by X-rays. *Aloe vera* is used in health drinks and a range of products now in the market.

Farmers have started cultivation of *Aloe vera* as it offers higher economic returns and grows in areas deficient in water. Planting can be done in rows 1.5 ft apart with distance between the rows being 2 ft. The varieties cultivated are *Aloe vera* or *Aloe barbadensis* Miller.

Aloe vera leaves contain juice around 75-80 per cent on a fresh weight basis. Plant is grown in the dry climate in presence of N-P-K fertiliser. It cannot grow under cold conditions and water near its root is not recommended. Production of *Aloe vera* plants is 15000 kgs. to 35000 kgs. per acre per year. The crop has a lifespan of five years when plants are attacked by microbes and quality of the leaves detoriorate.

Collected leaves are taken and sized into pieces which are passed through the toothed roller press made of mild steel. *Aloe vera* juice is extracted from the plant at the rate of 400 ml-600 ml from 1kg fresh plant's leaves.

Aloe vera leaves are harvested once in a year for the collection of juice. After collection of juice, leaves are used for manufacturing gel.

Take the leaves after extraction of juice and treat them with water at 50°C-60°C for two hr. After completion of extraction, liquid extracts are collected in a drum. Collected liquid is vacuum evaporated and concentrated to form a gel. After preparation of gel, it is packed in air tight plastic containers; 200 P.P.M. of methyl parabenzone is added as preservative for long storage.

Gel and Powder

Equipment

1.	Toothed Roller Press (M.S./S.S. Rollers) operated by 1 HP Motor	0.75 lakhs
2.	Extractor	0.75 lakhs
3.	Filter	0.50 lakhs

4. Vacuum evaporator	2.50 lakhs
5. S.S. grinder	0.50 lakh
	Rs. 5.50 lakhs

Yield	15000 kg.–35000 kg./acre
Gel per cent	400-600 gm/kg of leaves
Sales	6000 kg–9000 kg.
	@ Rs.100/kg.

Chapter 5

Rural Energy

Cultivation of *Jatropha curcus* and Production of Biodiesel

The oil bearing plant *Jatropha curcas* (L) (Jatropha) or physic nut is a drought resistant multipurpose large shrub or small tree not browsed by cattle. Although a native of tropical America it now thrives throughout Africa and Asia. It grows in a large number of climatic zones in tropical and sub tropical regions of the world and can be grown in areas of low rainfall and problematic soils. Jatropha is easy to establish, grows relatively quickly and is hardy. Being drought tolerant, it can be used to reclaim eroded areas, be grown as a boundary fence or live hedge in the arid/semi arid areas.

The wood and fruit of Jatropha can be used for numerous purposes including as fuel. Seeds of Jatropha contain (35 per cent by weight) viscous oil, which can be used for manufacture of candles and soap, in the cosmetics industry, for cooking or as a diesel/paraffin substitute or extender in the paint industry. This latter use has important implications of meeting the demand for rural energy and also exploring practical substitutes for fossil fuels to counter green house gas accumulation in the atmosphere.

These characteristics along with its versatility make it of vital importance to developing countries experiencing decreasing tree cover and soil fertility because of increasing population and development pressures. Nearly half the world's poorest people live on marginal lands with the number expected to increase from the present 500 million

Biodiesel is likely to emerge as an alternative fuel that could meet substantial oil need in the country. Oilseeds primarily used for developing biodiesel include *Jatropha curcas, Madhuca latifola* (Mahua), *Salvadora persica* (Pilu) *and Pongamia pinnata* (Karanja). The plants are found in abundance in States like Madhya Pradesh, Andhra Pradesh and Gujarat. If we use 20 per cent of non-forest wasteland in the country for the cultivation of certain types of oil seed plants, substantial demand of country's fuel needs could be fulfilled. Biodiesel is defined as monoalkyl esters of fatty acids derived from vegetable oils and animal fats. Biodiesel is obtained by chemically reacting vegetable oil/animal fat with alcohol to produce new compound, which is known as fatty acid alkyl ester. A catalyst such as sodium or potassium hydroxide is used and glycerol is produced as a byproduct with the following product:

$$\text{100 litres of Oil} + \text{10 litres of Methanol} \xrightarrow{\text{NaoH/KOH}} \text{90 litres of Biodiesel} + \text{10 litres of Glycerol}$$

(Menthol recovered)

Biodisel has been found to be non-polluting fuel with low sulphur content which is very good for engines. The exhausts from biodiesel are lower than the emission from regular diesel engines. Biodiesel provides substantial reduction in carbon monoxide, unburnt hydrocarbons and particulate emissions from diesel engines but there is slight increase of oxides of nitrogen (NO_x), which could be reduced by adjustment of engine's injection timing. Biodiesel has excellent lubricating properties. Even if 5-10 per cent addition is made to conventional diesel it can convert fuel with poor lubricating properties into an acceptable fuel. Biodiesel is also 'renewable' India has land mass of 329 million hectares out of which 65 million hectares is wasteland. These areas are isolated and fragile and suffer from prolonged draughts. Because of deforestation soils are susceptible to erosion resulting in decreasing top soil and ground water. These wastelands can be utilized for cultivating these sturdy plants for improving the economy of rural poor, get an alternative fuel which is less polluting and whose supply is not restricted like fossil fuel.

Agro-Technology

Soil

Wasteland unattractive for agriculture for low fertility and alkaline soils.

Rainfall

200 mm and 1200 mm in hot climatic conditions.

Planting

By seeds or rooted cuttings at a distance of 2m x 1.5m. 1600-2000 plants/ha or 400-500 plants/km.

Yield

0.4 T/ha/yr after two years and 2.5-3.5 T/ha/year after 5 years.

Return

3.00 T x Rs.5/kg–Rs. 15000/year for 20-25 years.

Plant Cost

Rs. 6.00 lakhs of 2001/batch

Low Cost Briquetted Fuel

In India at least 150 million tons of forestry wastes and 350 million tons of agricultural residue are available every year, which are not being utilized. A substantial portion of these residue goes waste and this even causes environmental pollution. At the outskirts of every Indian village heaps of agricultural wastes generate air pollution. On the other hand there is scarcity of fuel for domestic and commercial needs in the rural areas, which results in indiscriminate felling of trees in the forest areas. People, particularly women carry head loads of wood several kilometers to burn their chullas. It is possible to prevent or at least reduce the deforestation for fuel needs by providing alternative fuel as traditional fuel like coal, kerosene and gas etc. are beyond the reach of the people in rural or forest areas. Utilization of agro/forest wastes by conversion into fuel would help in resolving fuel problems.

Several technologies have been developed for conversion of biomass into fuel such as compaction of agro-waste into briquettes taking advantage of lignin present in the biomass as binder. But this technology has not become popular due to use of high pressure extruders/hydraulic presses, which requires high-energy input to operate these machines. Fluidized bed combustion/charring of biomass and briquetting the charred biomass using external binder is feasible.

Numerous charcoal production systems and improved techniques have been developed and are available for application from cottage level to commercial level using drum and mould methods to brick beehives, retorts and extruders. Appropriate process route has to be identified keeping in view the socio-economic and ecological factors.

Process

First of all the biomass is carbonized in a drum kiln under controlled supply of oxygen. The carbonized biomass is then mixed with 33 per cent (30-35 per cent) clay in presence of moisture. Black cotton soil is the best for this purpose. The mix is moulded using hand moulds similar to the one used for making bricks. After moulding, these bricks are dried in the sun and then used as smokeless fuel. Each briquette is of 175/180 mm dia with a thickness of 75/80 mm having 19 holes of 20/25mm dia. Weight of one briquette is about 300 gm which can be ignited in a ordinary clay chullah of round shape having two steel bars for supporting the briquettes. It lasts for 40-60 minutes for cooking purpose and about 1.5 hr for heating purpose and gives sustained and smokeless combustion. Each of the 19 holes act as a tiny gasifier and briquette once ignited provides nineteen distributed flames just like LPG stove. Two briquettes (one over other) can also be used if the *chullha* is made to accommodate them.

Calorific value of briquettes is 7000-7500 kcal/kg.

Moisture	7-8 per cent
Volatile Matter	13-14 per cent
Fixed carbon	80 per cent

Capacity

22,500 briquettes/year of standard size or 150 briquettes per day for 150 days per year.

Machinery Equipment

Following minimum equipment is required for a cottage scale unit:

Equipment	Number	Cost (Rs)
Drums	2	3,000
Stove	2	250
Moulds	2	1,000
Mixer/Pounder	1	2,000
Tools/Misc items		3,000
		9,250 or say Rs.10,000.00

Power Requirement

It is a hand operated unit and as such no electric power is required to operate the unit.

Raw Material

Biomass 150 kg/day for 150 days

Or

22,500 kg/year

Local clay–3000 kg/year

(20 kg per day for 150 days)

Others

Small quantities of water is required for moisturizing the mix.

Land		800 Sq.ft.
Low cost shed	250 Sq.ft.	Rs. 15,000
Machinery		Rs. 10,000
Working capital		Rs. 30,000
Profit		Rs. 30,000 per year

In south India, instead of clay, cassava or starch (5 to 6 per cent) is used binder and press could be used for making charcoal briquettes. This charcoal has several advantages over hard wood charcoals and used for industrial applications such as manufacture of carbon di-sulphide, carbon electrodes, carbon tetrachloride, sodium cyanide, activated charcoal, industrial ovens, water purifiers, and electrodes for primary batteries. Entrepreneurs can contact AMM Murragappa Chettiar Research Centre, Taramani, Chennai.

Solar Energy

With the global depletion of fossil fuels the need for alternative energy sources has become of prime importance. Among the possible alternative sources, the pollution free limitless source is solar energy. Considerable research and development work for thermal utilization of solar energy has been carried out all over the country to develop solar water heaters, solar space heating and cooling systems, solar refrigeration, solar pumps, solar power generation systems etc. At heart of all these systems is the solar collector. A large number of hot water solar flat plate collector systems have been already installed in the country.

Flat Plate Solar Collector

Several types of collectors have been developed. Among these one with copper tube and an aluminium sheet, and another with a copper tube and a copper sheet absorber were found most suitable. A flat-plate collector consists basically of a tube and a sheet absorber coated with a black paint and kept in an insulated box. The box is sealed with a glass sheet on the top. Water enters the collector through a header tube and passes through the riser tube and comes out from another header tube. Normally 50 liters of water can be heated upto 60° C on a bright sunny day using 01m^2 area of the flat–plate collector. It is, however, possible to heat the water to even 85° C using a flat–plate collector with selectively coated absorber. These flat-plate collectors with pipe line and storage tank are available in the market at a cost of Rs.20,000-25,000/m^2 area.

Now a days chinese solar water heaters using concentric glass tubes are available in market. Inner tube is coated with black selective coating and gap between the tubes is having vacuum to avoid radiation of heat.

Solar Cookers

Solar cookers of paraboloid stainless sheet were developed and used for making chapattis, frying, cooking etc. Subsequently, paraboloidal glass silvered at the rear side with a protective coating over the silvered side were developed. This has not only improved the life of the reflecting surface, but also increased its efficiency. A paraboloid cooker of one meter dia gives 1 KW of energy at its focus. In one of the training centre run by an NGO, meals for 40 persons are prepared by using solar energy.

Solar Oven

Another model with a plane booster mirror has been fabricated. This model is provided with an octagonal arrangement of booster mirrors to enhance concentration ratio and, temperature up to 250°C is achieved. It is used as an oven for use in domestic bakery. A cake is prepared in 40-45 minutes. Such ovens would be very appropriate for cottage scale bakeries in rural/semiurban areas. Hot-box type cooker with a single booster mirror has been widely accepted and manufactured in the country by various agencies. There is need to incorporate black nickel selective coating and anti-reflection coating over glass cover to attain maximum absorption developed by CECRI. Karaikudi. The effeciency of the present box type solar cooker could be increased by 40 per cent. Technology for black nickel coating is available from CECRI, Karaikudi and NAL, Bangalore.

Solar Drier

Sun-Drying is perhaps the oldest method of food preservation. It is still practised in the same manner as in ancient times. Sun-drying, however, has several drawbacks: it is slow, requires huge areas, and produces non-uniformly dried products. Moreover, prolonged exposure of the harvested crop to atmosphere leads to contamination with dirt and dust, insect infestation, occasional drenching with dew or rain, general decay, and quality degradation.

Marketing requirements and the desired quality attributes demand a quicker and hygenic method of drying spices, fruits and vegetables. Solar dehydration, which combines sun-drying with temperature control and regulation of air flow, not only improves the quality of the sun-dried products but also increases the drying efficiency. Solar driers developed by the Regional Research Laboratory

(RRL), Jammu, are a real boon to farmers in remote villages where electricity and other forms of energy are not readily available. The driers maintain 20-25°C higher temperature above the ambient and also offer better hygienic conditions. In the drying of items like apricots and chillies, the drier reduces the drying time is reduced to just 3-4 days and yields a wholesome product with better appearance and taste and longer shelflife.

The drying trays are kept in such a way that minimum shadow is cast on the heat–absorbing area. Two dampers are provided at the back and front of the drier for the control of air circulation and temperature. This temperature gives a temperature rise of 20°C to 25°C above the ambient. Drier with a drying capacity of about 25 kg of raw chillies has been fabricated and installed at number of places in Jammu and Kashmir.

The other type of solar drier is based on induced draft. The salient feature of this drier is that it is fabricated out of aluminium sheets and fitted with corrugated aluminium foil to provide more heating area and better heat transfer. The corrugated foil, which acts as fins is painted with a special black paint and the whole structure is housed in a wooden frame to give strength and to act as an insulator. The top is fitted with 4 mm thick glass panes and is connected to a separate drying zone. The drying zone, or the drying chamber is made out of wood and fitted with glass panes on three sides and also on the top, while the fourth side is provided with fixed racks over which wiremesh trays are kept. These trays are filled with the material to be dried. There is a chimney of suitable height. This creates a draft in the chamber, which in turn sucks hot air from the collector and becomes a continuous source of heating and drying.

These driers have been installed at several places.

Multi-fuel Cook Stoves

For cooking food, the rural poor use solid fuel and agriculture wastes, *e.g.* firewood, chips, leaf, twig, bush, husk, bagasse, cowdung cake, charcoal or coal briquettes depending upon availability at the point of time. Conventional country chulha (cook stove), besides having high fuel consumption causes smoke and related health hazards.

R and D work at RRL, Bhubaneswar and other institutes has led to the development of a series of low cost energy efficient multi-fuel chulhas which are nearly smokeless. The use of mix chulhas is being propagated by Ministry of Non-conventional Energy and are available from Sources at subsidised rates. The construction is simple and they are maintenance free. Over 2 lakh improved chulhas are being adopted every year in different rural areas. For this work the laboratory received the Pam Daver Basu Award 1994 for exemplary work in Rural Technology for homes/field/villages. Various models developed for the purpose are.

Harsha

Portable model: Fuel-coal, wood, twigs, dungcake, agro-wastes; thermal efficiency for coal 24 per cent wood 31 per cent, bagsse 34 per cent and, dungcake 43 per cent; fuel conservation 50 per cent over conventional; CO/CO_2 1.6–2.5; useful power rating: 1.5 kW: A larger size chulha has also been developed for community level applications.

Jyotsana

Portable model: Fuel-coal, wood and agro-wastes; thermal efficiency similar to Harsha model; circular configuration with low weight.

SK-1, SK-2, SK-3, SK-4

Stationery models with chimney: Fuel; wood, dungcake, agro wastes; thermal efficiency: 24 per cent to 40 per cent continuous burning, flexible dimensions.

SK–Delux

Stationary model without chimney: fuel, coal, wood, twigs, dungcake, agrowastes; thermal efficiency, for coal 24 per cent, wood 31 per cent, bagasse 32 per cent, dungcake 38 per cent; fuel conversation: 40-50 per cent, continuous ash disposal.

SK-Agni

Stationary model without chimney, lined-ceramic, mud-clad; fuel; wood, twigs, agro-waste; thermal efficiency: 29 per cent, fuel conservation; 50 per cent. It is the cheapest model.

Purification of Water by Solar Energy

The current boom in population in urban areas and the increasing intensive irrigation requirements for agriculture in rural regions have already resulted in considerable drying up of wells and aquifers. The decreasing availability of water has necessitated in the search for fresh sources of drinking water. The available water in many areas in the country is brackish, saline or impure. Though efforts are made to supply water through public distribution systems there are inherent limitations to this programme. This is where the (Murugappa Chetiar Research Centre) principle of augmentation of water availability limited to individual/groups of households becomes pertinent. This is easily applicable in rural and urban India alike as the energy requirement for the distillation units is met from solar radiation. India is gifted with rich sunshine and therefore solar energy is not limit here. MCRC's involvement in fabrication of solar stills started in 1979 with the basic conical type named "Thoyam". In Thoyam a pan of impure water is enclosed in a transparent cone-shaped cover and traps solar heat causing part of the heated water to evaporate, rise to the top and condense on the cooler transparent surface inside the cover. The pure water thus formed drips through a channel to a reservoir. This set up fabricated from inexpensive and locally available materials, can produce 3 to 4 litres/m^2/day in bright sun light. Field trials in the MCRC has designed new solar stills through its RandD processes to yield 8 to 10 litres of potable water/m^2/day, which is sufficient for a family.

Biomass Gasifier

Biomass is basically an organic material which stores solar energy from sun through the process of photosynthesis. Biomass fuels include agricultural waste, crop residues, fuel wood, woody waste and organic waste like tapioca stalks, soyabean stalks, ground nut shell, coconut shells, sugar cane trash and corn stalks/cobs. Biomass does not release CO_2 to the atmosphere unlike fossil fuels as it absorbs the same amount of carbon during growth. It is appropriate, cheap, eco-friendly and renewable source of energy.

Biomass gasifier converts solid biomass into more convenient gaseous form. Normally, the ratio of air to fuel required for complete combustion of the biomass is 6:1 to 6.5:1 which is defined as stoichiometric combustion with end products being CO_2 and H_2O. In the gasifier the air-fuel ratio is 1.5: 1 to 1.8:1 under sub stoichiometric conditions. The gas produced is producer gas which is combustible having calorific value of 4.5-5.0 MJ/Kg.

The gasifier system comprises of a reactor for generation of combustible producer gas which after cleaning and cooling process is made available for power generation/thermal applications. The producer gas from gassifier can be used to run diesel engine. Gasifier when connected with electric generator through engine can generate electricity. One cu.m. of gas can generate thermal energy of the order of 4.5-5.0 MJ with flame temperature of upto 1200° C.

Wind Power

Wind energy is created due to heating of Earth's surface and the rotation of Earth. Uneven heating of different parts of the earth causes difference in the air pressure causing flow of air from high pressure zone to low pressure zone. Wind contains tremendous amount of energy (whirlwind causes havoc in the area) which is now used to generate power on large scale. To set up wind mills windy areas are identified and wind flow data is compiled for economic viability of the system. The wind powder is used to generate mechanical power or electricity. Wind turbines convert the kinetic energy of the wind into mechanical power which can be used to accomplish specific tasks like grinding grain, pumping water or to operate electrical generators. In country like Switzerland wind power is generated on a large scale. High wind velocities are available in high altitude or coastal areas. Wind Energy systems are also working in some parts of out country. The wind turns the blades (like, air craft propeller blades) which spin a shaft connected to a generator and makes electricity.

Modern wind turbines fall into two basic categories: the horizontal- axis like the traditional farm wind mills used for pumping water and the vertical-axis design like egg beater style Durries model, (named after French inventor). Most large modern wind mills are of horizontal-axis.

Components of Horizontal Turbines Type

- ☆ Blade or rotor which converts the energy in the wind for rotation of shaft.
- ☆ A drive train-gear box and a generator.
- ☆ A tower that supports the rotor drive train.

☆ Auxiliary equipment like control system, electrical cables, ground support and inter connection equipment.

Usually the utility scale turbines in the range of 50–750 KW for homes, telecom dishes, water pumping etc. are available from manufacturers.

Solar Photovoltaic Cell System

In solar photovoltaic system (SPV), electricity is generated from sun rays through semi-conducting solar cells which convert heat energy into electrical energy. SPV technology has immense potential for generation of electricity for applications such as solar lanterns, solar street lights, domestic lights, water pumping system, and solar power generation.

SPV are modular in nature and they could be connected like battery to enhance power capacity. They are eco-friendly as they do not consume any fossil fuel and are easy to install, operate and maintain.

A compact lighting system consists of a lamp (7-8 W), battery power pack and electronic circuitry housed in metal, plastic FRP casing and a PV module. The lantern can operate for 3-4 hrs once charged for 5 hrs from sun.

Solar Home Light system is aimed to provide electricity for operating lights and/or fans or portable TV sets for a few hours a day. A typical system consists of a 37 W Solar PV module, a 12 V 40 Ah low maintenance tubular battery and control system. The invertor operates for 4 hours when charged during the day.

Biogas Plant

First biogas plant was developed by ICAR in 1940 for generation of fuel gas and bio-manure to meet the growing needs of rural population. The spent slurry is a rich source of plant nutrients and enhances soil productivity and quality of organic food. There are about 4.0 million biogas plants in the country which help in self–sustainable agricultural production and reduces the cost of farm inputs apart from methane rich biogas. The Biogas management nourishes the soil with supply of essential nutrients to accelerate the growth of plants especially for root growth. It enhances the aeration of soil for root penetration resulting in better growth.

Consumption of weed killers is reduced and water holding capacity of the soil is enhanced ensuring plant growth. In general the soil health improves.

Types of Biogas Plants

Floating gas holder Type (KVIC model) consists of mild steel gas-storage drum, inverted over the digester and moves up and down around a guide pipe corresponding to the formation and utilization of gas.

Fixed Dome Type: In this model, the digester and holder are part of composite structure made of cement and brick masonry.

Two designs are available:

Janta model: Consists of cylindrical digester with a dome-shaped roof with large inlet and outlet tanks. It costs 20-30 per cent less than the cost of the floating type drum.

Deshbandhu Model: Consists of main tank comprising two segments; one for digester and other for storage of gas in the same sphere. This model is 20 per cent cheaper than Janta model and nearly 40 per cent cheaper than floating type model.

The plants should be cured by putting wet gunny bags over the dome and the inlet and outlet tanks. Curing with water inside the dome is also carried out for 3 weeks.

Dung Requirement

Plant Size cum	*Dung/day in kgs*	*Cooking Gas for Number of House Hold Members*
1	25	3-4
2	50	5-8
3	75	8-12
4	100	12-16

The cost of 3 cum plan is about Rs. 12-15,000. The pay back period is 2-3 years keeping in view the cost of gas and bio-manure obtained.

The use of bio-manure and bio pesticide from cow urine and extraction of neem leaves/kernels is carried out in organic farming.

Chapter 6

Minor Forest Produce or Non-Timber Forest Produce

About 21 per cent area of the country is covered under forests. Exploitation of Non-Timber forest produce offer scope for development of small/cottage scale industries and value addition of this resource can offer jobs to millions in rural and tribal areas and improve economy of these backward areas. A calendar can be drawn for engaging the rural/tribal population for collection of produce for atleast 100-150 days in a year. Various Research and Development (RandD) Organisations of the country have developed technologies and processes for value addition of MFP/NTF produce. Some examples have been given in the Table.

Table: Industrial Uses of Medicinal, Aromatic and Other Economic Plants

Sl.No.	Name of the Plant	Chemical Constituents and Plant Part(s)	Use	Availability of Technology	Remarks
(I) MEDICINAL PLANTS					
1.	*Terminalia chebula* HARRA	✰ Tannins 20-40 per cent, astringents substances 30 per cent, 18 amino acids, glucose, sorbitol, etc. ✰ Fruit	Ayurvedic medicine	Technology is simple and available with any ayurvedic pharmacy.	The fruits are used singly or in mixture in various ayurvedic formulations and have good demand.
2.	*Terminalia bellerica* BAHERA	✰ Tannins, β-sitosterol, gallic acid, etc. ✰ Fruit	-do-	-do-	-do-
3.	*Terminalia embilica* AMLA	✰ Vitamin C, etc. ✰ Fruit	-do-	-do-	-do-
4.	*Terminalia arjuna* ARJUN	✰ Oxalic acid about 20 per cent ✰ Bark	Used in various pharmaceutical and chemical industries	Technology available with NRDC, New Delhi.	—

Contd...

Table–Contd...

Sl.No.	Name of the Plant	Chemical Constituents and Plant Part(s)	Use	Availability of Technology	Remarks
5.	*Holarrhena antidysenterica* KURCHI	☆ Kurchicine and other alkaloids ☆ Bark	(*i*) Ayurvedic medicine as 'kurachal' (*ii*) Tincture or extract of bark containing kurchicine, used in modern system of medicine as an anti-dysenteric drug	Technology can be made available from CIMAP, Lucknow.	Wood used for making toys.
6.	*Adhatoda zeylanica* ADUSA	☆ Vascicine ☆ Leaves	Used in modern system of medicine as oxytocic	Technology can be made available from RRL, Jammu.	
7.	*Semecarpus anacardium* BHILAWA, MARKING NUT	☆ Semi-drying oil- 20-28 per cent ☆ Nut- 32-35 per cent phenol.	Used in indigenous systems of medicine.	Oil can be extracted by simple technology IICT Hyderabad	Good demand of nuts and oil in the market.
8.	Van Tulsi	☆ Leaves	Essential oil	-CIMAP (LKO)	Demand in Pharma Industry
(II) AROMATIC PLANTS					
1.	*Vetiveria zizaniodes* KHUS	☆ Essential oil 0.3 to 0.7 per cent. ☆ Root	Perfumery and cosmetic industries	Distillation technology available with CIMAP, Lucknow	The oil has good demand both within and outside the country.

Contd...

Table–Contd...

Sl.No.	Name of the Plant	Chemical Constituents and Plant Part(s)	Use	Availability of Technology	Remarks
2.	*Cyperus scariosus* NAGARMOTHA	☆ Essential oil 0.2 to 0.4 per cent ☆ Root	Perfumery and cosmetic industries	Distillation technology available with CIMAP, Lucknow	3 plants in production.
(III) OILS AND FATS					
1.	*Madhuca longifolia* MADHU GULLI	☆ Yields oil ☆ Seed	Used in soap and other industries	Technology available with Indian Institute of Chemical Technology, Hyderabad, and Khadi and Village Industries Commission, Bombay	The seeds are collected by the tribals and sold in weekly markets.
2.	*Schleichera oleosa* KUSUM	☆ Yields oil ☆ Seed	Used in soap and textile industries	-do-	-do-
3.	*Mangifera indica* WILD MANGO	☆ Yields oil ☆ Ker nel	Oil is a substitute for cocoa butter	Technology can be made available from National Institute of Nutrition, Hyderabad	
4.	*Soap Nut* REETHA	☆ Saponin	Shampoo Ind.	RRL Bhub	Available in Betul
5.	*Shikakai*	☆ Cleanser	Shampoo Ind.	-do-	Available in Betul

Contd...

Table–Contd...

Sl.No.	Name of the Plant	Chemical Constituents and Plant Part(s)	Use	Availability of Technology	Remarks
6.	Black Berry JAMUN	☆ Yield of pulp seeds	Vinegar and dry seed for antidiabetic uses.	Author	
7.	*Shorea robusta* SAL	☆ Yields oil ☆ Seed	Oil is a substitute for cocoa butter	National Institute of Nutrition, Hyderabad	
(IV) STARCH-YIELDING PLANTS					
1.	*Mangifera Indica* WILD MANGO	☆ Yields starch ☆ Ker nel	As starch	Technology can be made available from CFTRI, Mysore	Starch can be used in textile and other industries. The by-product, mango peels (dried slices of unripe mangoes) has good market and demand.
2.	*Curcuma angusti-folia* TIKHUR	☆ Contains edible starch ☆ Tuberous root	High quality starch used in milk puddings, etc. and sold in metropolitan cities at a fairly good price	CFTRI, Mysore can help in making the technology available.	The tribals of Bastar district extract starch from the roots by crude methods.

Contd...

Table–Contd...

Sl.No.	Name of the Plant	Chemical Constituents and Plant Part(s)	Use	Availability of Technology	Remarks
(V) LIQUOR INDUSTRY					
1.	*Madhuca longifolia* MAHUA	☆ Corolla contains protein 4 per cent fat 0.5 per cent and total sugars 72.9 per cent ☆ Flower	Used in preparation of liquors Honey like syrup.	CFTRI, Mysore has developed a method of preparation of brandy from mahua.	Mahua is sold in the weekly bazars by the tribals.
(VI) EDIBLES					
1.	*Buchanania lanzan* CHIRONJI	☆ Kernel edible	Used as 'mewa' or desert.	Technology needed only for decortication by a sophisticated machine and can be made available by CFTRI, Mysore.	Chironji is sold in the weekly bazars by the tribals. There is good demand for roasted kernels as deserts in foreign countries.

Contd...

Table–Contd...

Sl.No.	Name of the Plant	Chemical Constituents and Plant Part(s)	Use	Availability of Technology	Remarks
(VII) TANNINS					
1.	*Terminalia chebula* HARRA	☆ Contains tannins 20-40 per cent ☆ Fruit.	Largely used in leather industry and in manufacture of ink, textile stamping, dyeing and pharmaceutical industries. Gallic acid.	Technology available with CLRI, Chennai.	Annual production of tannins in the country is 50,000 tonnes as extracts. Apart from this, India imports tannins worth about Rs. 4.72 crores. Drug intermediate 3-4-5 TMBA.
2.	*Cleistanthus collinus* GARARI	☆ Tannins in fruit-14 per cent, twing bark-21 per cent, leaves 19 per cent and stem bark-33 per cent	-do-	-do-	
(VIII) NATURAL DYES					
1.	*Woodfordia fruticosa* DHAWAI	☆ Vegetable dyes ☆ Flowers	Vegetable dyes are used in textile, indigenous pharmaceuticals and liquor industries.	Technology can be had from NRDC, New Delhi.	Flowers are sold by the tribals in weekly bazars.
2.	*Butea monosperma* PALAS	☆ Food colour	-do-	CIMAP (Luck)	-do-

Contd...

Table–Contd...

Sl.No.	*Name of the Plant*	*Chemical Constituents and Plant Part(s)*	*Use*	*Availability of Technology*	*Remarks*
3.	*Bamboo dandro calamus*	—	Agarbatti sticks and furniture, handicrafts	Processing machines are available	
4.	*Sisal* Fibre	—	Ropes and Handicrafts Waste leaves for hand made paper.	RRL Bhopal	
(IX) GUMS AND RESINS					
1.	*Stercula urens* GUM KARAYA	☆ Stem exudes gum	Gums are used in various industries like confectionery, toltery, sizing and calico-printing, etc. It has also good demand in foreign countries.	Collection, grading, and packing, needs to be done in a scientific manner for high quality gums.	Tribal co-operative societies may be contacted for collection and sale of gums. TRIFEED can help IGRRI Ranchi.
2.	*Shorea robusta* SAL GUM	-do-			
3.	*Anogeissus latifolia* GUM GHATTI	-do-			
4.	*Pterocarpus marsupium* KINO GUM, etc.	-do-			

Contd...

Table–Contd...

Sl.No.	*Name of the Plant*	*Chemical Constituents and Plant Part(s)*	*Use*	*Availability of Technology*	*Remarks*
5.	SEED GUM *Tamarindus indica* IMLI	☆ Poly saccharide composed of xylose, glucose, galactose etc.	Tamarind kernel powder (TKP) used in textile industry, in sizing and match industries, etc.	Technology can be made available from CFTRI, Mysore and NBRI, Lucknow.	TKP processing units can be established at cottage level.
6.	Honey	—	Purification unit	Bee and Honey Research Institute, Pune	Potential Export
(X) BROOMS					
1.	*Thysanolaena maxima* PHOOL BAHARI PHOOL JHADU	☆ Inflorescence used as broom	Brooms are used throughout the country.	Technology simple and well known	Narainpur tehsil of Bastar yields the best phool baharis in the country.

Process Flow Sheets

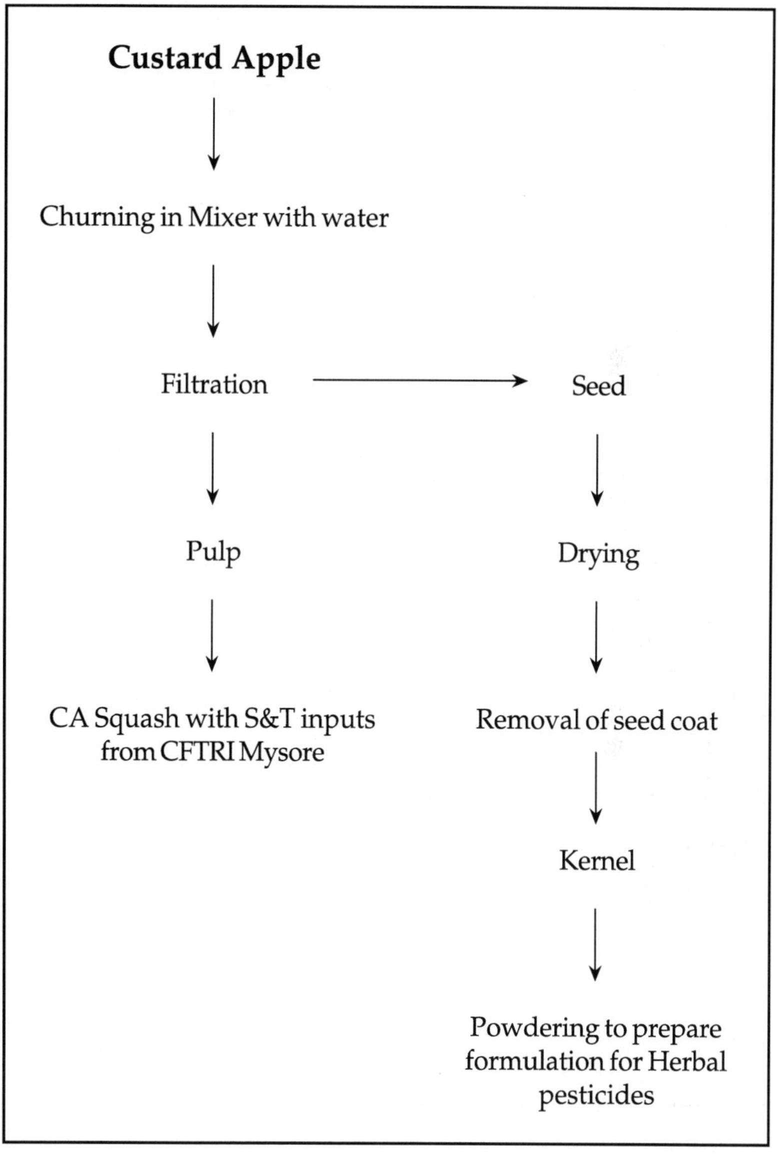

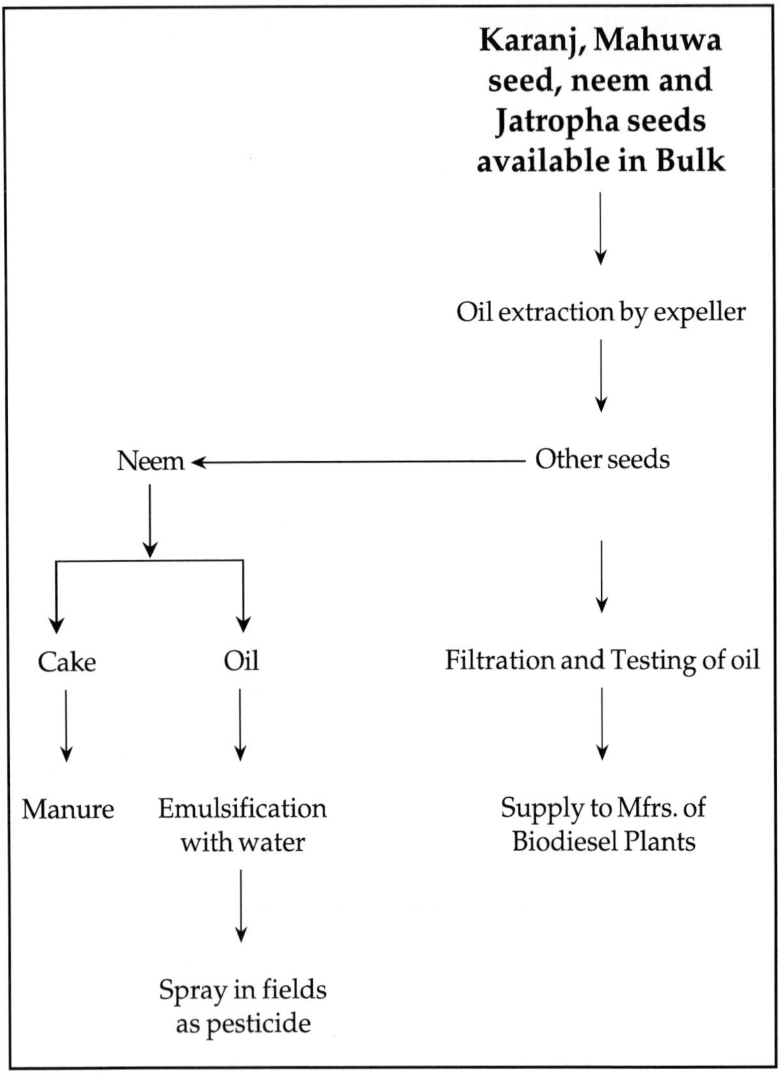
Karanj, Mahuwa seed, neem and Jatropha seeds available in Bulk
Oil extraction by expeller
Other seeds
Neem
Cake
Oil
Filtration and Testing of oil
Manure
Emulsification with water
Supply to Mfrs. of Biodiesel Plants
Spray in fields as pesticide

Palmarosa, Tulsi (*Ocimum*)

↓

Extraction of essential oil using existing distillation Plant

↓

Scientific Storage

↓

(Gas liquid chromatography)

↓

Supply to Market

Harra, Bahera and Aonla

Drying

↓

Decortication of species → Seed

↓

Powdering

↓

Packaging in Air Tight Bottles

Chironji

↓

Decortication

↓

Separation seed–Husk

↓

Grading

↓

S&T input for Treatment to enhance shelf-life

↓

Packaging in attractive packs

↓

Marketing

Bhilawa (Seme-carpus)

↓

Contains 32-35 per cent phenol

↓

Extraction of oil using extractor

↓

Filtration

↓

Reaction of oil containing phenol with formaldehyde (phenol + Formaldehyde)

↓

Phenol Formaldehyde Resin

↓

Used as binder in wood working and plywood industry

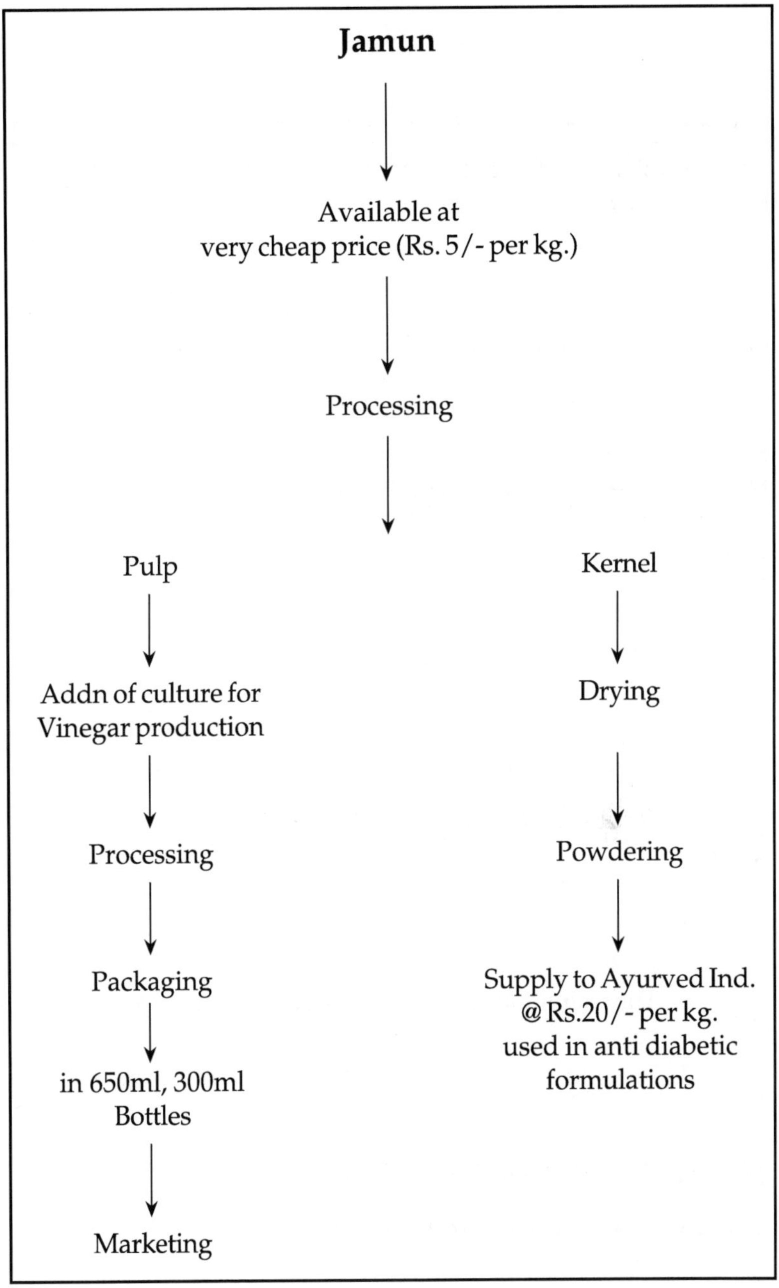
Jamun
Available at
very cheap price (Rs. 5/- per kg.)
Processing
Pulp
Addn of culture for
Vinegar production
Processing
Packaging
in 650ml, 300ml
Bottles
Marketing
Kernel
Drying
Powdering
Supply to Ayurved Ind.
@ Rs.20/- per kg.
used in anti diabetic
formulations

Honey Collection and Processing

Honey has become part of diet of a large segment of the population not only in India but also in the world. Apart from the nutritional value it has medicinal properties as well. The Rock Bee is the largest producer of honey among Indian Bees. One comb yields 3 to 15 kg of honey. But the beehives are formed at places which are difficult to approach.

Crude methods of collecting honey adopted by tribal honey hunters using fire and squeezing of combs by hands leads to destruction of the whole colony of bees and recovery of honey is only 30pc. Loss of bee population hampers cross pollination. The traditional honey hunters face considerable hazards in their vocation due to the inaccessibility of honeycombs and the ferocity of these wild bees. New techniques have to be propagated for collecting honey which is more safe and ensure more profits than the existing ones. Honey hunting is being thought of as an additional source of income for the villagers.

Advantages of the New Technique

The improved technology of honey extraction has several advantages.

1. Honey is obtained without destruction of the comb or bees.
2. Makes available the better quality of honey for the consumers.
3. Availability of honey collection increases.
4. Safety of honey collectors is ensured.
5. Preserves the bee population, ensure multiple yields and promotes cross pollination.
6. The technique multiplies number of extractions from each colony.
7. Provides higher incomes to the honey gatherers and their families due to enhanced production.

Improved Technique

The technology package comprises:

1. *Water Spray*: This is done to overcome the ferocity of the bees and facilitate their handling by honey hunters. Plain

water is sprayed over the rock bee colony that serves to restrict the free movement of bees by wetting their wings, and lowers the temperature of the colony. A bucket spray pump and hose are used: one person pumps water from the bucket kept at ground level and another directs the hose pipe with a fine spray nozzle at the comb.

2. *Protective Dress*: Full body protective dress, made of double layered thick yarn cloth, with suitable facial protection provides complete safety against any possible poisonous bee-sting. A pair of gloves and suitable footwear are also used.
3. *Portable Rope Ladder*: This is used by the honey hunters to reach bee colonies, at high and inaccessible rock. The 90 ft. ladder (weighing about 18 kg) uses nylon ropes, 90 rungs made of bamboo lengths and reinforced by steel wire. An iron ball attached to one end of the rope is thrown across the branch to which access is desired and the ladder is pulled up.
4. *Clip*: A simple contraption is used to hold the brood chamber of the colony and prevent it from falling when the honey chamber is removed, thus enabling the bees to rapidly rebuild their colony which in turn facilitate repeated honey extraction. A clip made of two bamboo splits, attached to the branch by the wires at one end and joined by a hinge at the other, is used to clasp the comb from both sides at the line between the brood chamber (lower) and the honey chamber (upper). Since both ends of the clips are fixed to the branch, the colony continue to work normally and rebuild the honey chamber for honey storage.
5. After wearing sting proof dress, 2-3 members of the team climb the tree with the help of a ladder provided to them. Then every approachable bee colony is operated separately by gently cutting off only the 'Major honey containing portion' from the hive without harming the rest of the comb.
6. *Filtration and Refining*: After collection the crude honey is transferred to the double walled S.S. jacketed vessel provided with heater at the bottom. The water in the outer jacket is slowly heated upto 60° C which slowly provides

heat to honey kept in the inner vessel to reduce its viscocity Thereafter the honey is transferred to screw press for pressure filtration. This removes waxes sting and other extraneous matter. The honey is stored in SS container.

7. The materials used for the extraction of honey include torch, thin rope, belt, basket and a special type of knife.

Process Flow Sheet

Scientific Collection

↓

Heating (Multi Jacket Chamber)

↓

Removal of Scum

↓

Storage

↓

Filtration (Pressure filter)

↓

Storage

↓

Bottling

↓

Packing

Source of Technology: Centre of Source for Villages (CSV), Sewa Gram, Wardha.

Techno-economics

Capacity	2000 kg.	
Work shed 600 sq.ft.	1,20,000	Rs.
Machinery/equipment	30,000	
Working capital	50,000	
	2,00,000	
Sales		
Honey 2000 kg x 120	2,40,000	
Wax 200 kg.	20,000	
	2,60,000	
Profit	60,000	Rs.

Lac Processing

Lac is a gift of nature and India is especially endowed with lac which is a minor forest produce. The description of lac is found in Atharvaveda wherein it has been mentioned as 'Laksha'. Lac is a resinous material produced by a kind of insect on branches of some particular varieties of trees known as lac host trees. The main host trees are kusum, ber, khair, bhalia and palash. The sticklac collected from trees contain lac, dead lac insect, bark leaves of trees, earth and sand. Besides these impurities, lac contains a dye having body fluid of insect. Before 1900 AD the sticklac used to be processed for collection of dye, for use in printing of textile meant for national and international markets and also adhesives and commercial ink. But with the advent of synthetic dyes the demand for natural dyes has reduced; the demand curve is rising again due to toxic effects of synthetic dyes.

India is the largest producer of lac and its share is more than 50 per cent. About 98 per cent of lac cultivation is concentrated in Jharkhand (Chhota Nagpur), M.P., W.B. and U.P. It is a source of livelihood for three million tribals and other rural population. There is export of lac to the tune of Rs.100 crores to nearly 100 countries.

Indian Lac Research Institute, now India Resins and Gums Research Institute Nam kum, Ranchi, has made significant contribution in development of cropping practices and development of new products for diverse applications in Industry. The institute conducts short term as well as long term training programmes for farmers and entrepreneurs. It also renders advisory services to solve technological problems.

Advice is rendered for following operations:

- Pruning techniques for lac hosts
- Spacing system: kusum, pales and ber.
- Air laying techniques–palas–kusum and galwang
- Intercropping: colocasia, turmeric and Dinanath grass complete system.
- Various aspects of crop inoculation.
- Pest management.
- Harvesting
- Storage and transportation of brood lac.

Chapter 7

Purification of Drinking Water

Life on this planet has been possible due to availability of water which is the basic necessity of life. Two third of the universe is covered with water. Water in human body is in the range of 60-70 per cent which is necessary to regulate various functions in the body such as providing food and nutrition to all organs, urination, excretion and maintaining level of heat. Water is also necessary for taking bath, washing clothes, cleaning of premises, building construction, irrigation and industry etc. For drinking purposes water should be clean, colourless, devoid of smell, sweet and tasty and free from bacteria. At some places water is not potable due to dissolved minerals beyond permissible limits such as flourides, iron, lead, arsenic, manganese, nitrates etc. It has been found that in about 230000 villages of the country water is not available or not potable due to impurities. W.H.O. has reported that 80 per cent diseases are caused due to unclean or shortage of water. It is estimated that 50 per cent doctor's bill could be saved if pure drinking water is provided in the villages.

Water pollution is also caused by human beings and due to industrial effluents which give rise to microbes which pollute water with poisonous chemicals and population suffer due to water borne diseases like diarrhpea, cholera, jaundice, typhoid, etc. Mosquitoes, flies and various insects grow on stagnant water which cause malaria and skin diseases. Availability of pure drinking water, therefore, is a necessity of life and fundamental right of every citizen. However, in remote areas where Government functionaries can not reach should purify water and save themselves from harmful diseases.

Several cost effective technologies are available for this purpose which can be adopted in rural and semi urban and urban areas.

Slow Sand Filtration Unit

Drinking water should be clear, tasteful and free from minerals and bacteria. Often these properties are not found in water from natural sources and it becomes essential to purify the water. Slow sand filtration unit is suitable for sources like canals, lakes, rivers and reservoirs. The design of slow sand filter is very simple and easy to construct and maintain. It enables recovery of water free from bacteria, dissolved solids, carbonic matter etc and provides water of uniform quality without using costly chemicals. This methods should be used when turbidity is below 30 N.T.U.

Before the use of slow sand filtration process water should be cleaned by decantation in a tank or filtered to achieve turbidity below 30 NTU. This method can be adopted in rural areas where chemicals, expertise and labour are not readily available but there is no dearth of land, labour and sand. In this process a 3 meter deep tank is constructed. At the bottom of tank boulders upto a height of 80-100 cm are placed. Then a 80-100 cm thick layer is filled with sand. Water after passing through layers of sand and boulders is collected at the bottom and flows to another tank for use.

Natural colour	30-100 per cent reduction in colour
Turbidity	1 NTU
E. coli	95-100 per cent less
Cist	Complete eradication
Carbonic matter	60-75 less.

For a village of 1000-1200 population two tanks of 2 sq.m. surface area and height of 3 meters are sufficient.

Iron Removal

Iron is an essential nutrient for human and animal life, and plants. It is found in nature in various types of soils, boulders, sand, rocks in the form of oxides, some times in the form of ferrous carbonate and sulphates. These salts are carried by rain water during percolation and mix with sub surface water. When the water is pumped out, due to oxidation the iron becomes available in insoluble or colloidal form resulting into browning, turbidity and impart metallic and bitter taste to the water. Tea made from iron contaminated water becomes black and clothes washed with this water become brownish. Distribution of iron contaminated water is problematic and could choke the pipeline. According to the Ministry of Health and B.I.S., water containing 0.3 mg. of iron per litre is acceptable but maximum permissible limit is 1 mg per litre. If contamination is beyond 0.2 mg/l it gives rise to formation of iron bacteria leading to iron diarrhoea and it is essential to make water iron free. National Environmental Engineering Research Institute, Nagpur has developed a kit to detect iron in water. 50 ml of water is filled in a tube. Break the ampule developed by NEERI and transfer the chemical to the nestler tube. Allow the chemicals to react for 10 minutes. If the column of tube becomes orange red, it confirms the presence of iron beyond permissible limit. Transfer the solution to the test tube of NEERI developed kit and compare with the chart to estimate percentage of iron.

Various technologies have been developed for iron removal but slow sand filtration method is simple and easy to operate. Water is allowed to pass through layers of coke, lime, lime stone, calcite etc. Iron in the form of ferrous (hydrous) sticks to the medium and one gets iron free water.

Alkaline process can also be adopted where lime and soda ash is available. Chlorination can eliminate iron bacteria but the process is applicable if quantity of CO_2 is low: In case quantity of CO_2 is more than iron, construction of oxidation tanks becomes necessary.

Chemical Process

Various types of iron contamination could be eradicated by the reaction of sodium carbonate and sodium phosphate with iron present in water. Contaminants in the form of precipitate settle at the bottom of the tank.

In Domestic Process take: 60 litre bucket and fill 40 litre water in it. Stir for 10 minutes, allow the impurities to settle. After two hours filter the water through sand bed in a second bucket. Thus pure and safe drinking water is obtained.

Iron Removal Plant Connected to Hand Pump

NEERI Nagpur and CMERI Durgapur have developed iron removal plants connected to India Mark II hand pumps which can be fabricated in rural areas using mild steel plate, pipe, valves and consumables like lime stone, manganese and sand. Fabrication workshops in rural area can easily fabricate the unit. Water is forced to the top of the plant when it percolates through layers of sand, stone and collected at the bottom. Water containing 1-5 mg/L of iron and 0.1–0.3 mg/L of manganese can be purified at the rate of 200 L/hour. As regards maintenance, sand and lime stone could be changed periodically when the plant gets clogged.

Chlorine Tablets

Human beings and animals are major causes of pollution which make water sources unfit for human consumption. In cities due to inadequate chlorination/purification of water people use water filters, acquaguards etc. but in villages such facilities are not available which gives rise to diseases like cholera, typhoid, jaundice etc. In this state of affairs people should use chlorine tablets and ampoules. For chlorination of water, a tablet is broken, added to water, stirred and left for 30 minutes to complete the reaction of chlorine, emanating from the tablet, with bacteria in the water. Subsequently, water is filtered for drinking purposes. One tablet costing Rs.0.10 is sufficient to purify 20L water which meets the need of a family of 4-5 members.

Pot Chlorination of Wells

In most of the villages main sources of drinking water are open wells which are contaminated by dry dust and dirt. In addition pulling ropes and unclean utensils cause water pollution which cause waterborne diseases. The traditional methods adopted to purify water have drawback like uneven release of bleaching powder causing excessive release of chlorine. NEERI, Nagpur has improved the process for cleaning of well water in which a plastic container with cover is used. In the lid of container two 0.5cm dia holes are made and stone pebbles of about 5cm. thickness are filled in this container upto a height of 20-25 cm. Bleaching powder and sand in the ratio of 1:4 are filled on the surface of pebbles and lid is tightly fixed. This plastic container encased in a cage is lowered in the well with the help of a string so as to submerge 1 metre deep in the water. Due to reaction of water with bleaching powder chlorine is emitted which disinfects the water for 10-15 days. The process can be repeated regularly. By using this process quantities of chlorine in the range of 0.5-0.2 Manganese per litre can be easily maintained. One kg of bleaching powder is sufficient for 1200 L/day discharge. Total outfit cost Rs.50/- and expenditure of Rs.30/- per family per year is very reasonable.

Solar Still

In many parts of country saline water is found in sub-surface sources due to percolation of salts in water lakes beneath the earth. Women bring sweet water from wells located long distances away from their homes and villages causing them drudgery.

Central Salt and Marine Chemicals Research Institute, Bhavnagar Gujrat has developed small solar stills to provide drinking water by purifying saline water and several such units have been set up in Gujarat. In this water is filled upto a depth of 5cm in still. Due to heat of the sun water evaporates and water vapours settle inside surface of glass cover and trickle down the slopy surface and are collected through a channel located at the end of glass cover. About 20-25 litres of water is obtained from one sq. meter of glass cover. Designs of solar stills of capacity 100–80,000 L/day are available from CSMCRI, Bhavnagar. Cost of a 100 L/day solar still is about Rs.5000/-.

Fluoride Removal

About 23,000 villages in India are affected due to the presence flouride in water. Flouride content of 1 mg per litre is permissible but higher content of fluoride causes deformation of bones, yellowing of teeth and aging. NEERI, Nagpur has developed a simple process for flouride removal beyond permissible limit called "Nalgonda process".

In the domestic process water is filled in a bucket of 20-25 litre capacity having a tap 3-5 cm above the bottom. Keeping in view the alkalinity of water lime and alum powder are added to water and stirred for 15-20 minutes; water is allowed to settle for one hour. Impurities settle at the bottom of the bucket. Thereafter clean water is taken out by decantation from the tap and impurities are thrown out and the bucket is cleaned for next day. 400 mg each of lime and alum are sufficient for one cu.m. of water having flouride content of 3 mg/ litre.

It is preferred to use solutions of lime and alum. To prepare 10 L solution of alum take 1000 gm alum in plastic can and slowly dissolve in water so as to make 10 L solution, one ml solution contains 100 mg alum. Similarily take 100 gm of slaked lime and add to water slowly to make 10 L of solution. 1 ml of this solution contain 10 mg of lime. Both lime and alum are available easily in rural markets. Sometimes small quantities of bleaching powder and sodium carbonate are also used in addition to lime and alum. The process is very economical and annual expenditure for a family of 5-6 person is Rs.15-20.

Chapter 8

Rain Water Harvesting

During rainy season substantial quantity of water run-off causing floods resulting in loss of property, human and animal population. This could be checked at least partially by the application of Rain Water Harvesting techniques of collection and storage of rain water at surface or in sub-surface aquifer for use during summer or period of scarcity. Artificial recharge of ground water augments the underground level of water at a rate exceeding natural conditions.

There is decrease of infiltration of rain water in the sub-soil due to urbanisation, increase in population and irrigation. Increasing vegetation cover helps to improve ecology of the area. The quality of ground water also improves by dilution due to percolation of rain water.

The methods used in urban areas and rural areas differ due to hydrogeology and method of usage. In urban and semi-urban areas Roof top rain water is recommended for recharge of pit and trench, tubewells or recharge wells. In rural areas rain water harvesting is done through Gully Plug, Contour Bund, Gabion Structure,

Percolation tank, Check Dam, cement plug, Nala bund, recharge shaft, Dugwell Recharge etc.

In roof top rain water harvesting the rain water is chanelled through the drain pipes and pass through beds of coarse sand (15 per cent), gravel (20 per cent) and boulders (50 per cent). The depth of this filter bed varies between 1.0-1.5 meters. The filtered water is utilised to re-charge tube wells or surface tanks. Generally rain water of first rain is not collected/harvested to prevent entry of dust etc. in to the recharge.

Gabion structure is a kind of check dam commonly constructed across small streams to conserve stream flows with practically no-submergence beyond stream course. A small bund across the stream is made by putting locally available boulders between mesh of steel wires anchored to the banks of stream. Such structure is recommended upto height of 0.5m on stream of 10m width.

Availability of Rain Water through Roof Top Rain Water Harvesting

	Rain fall	
Roof Top area Sqm.	*100mm*	*400mm*
20	1.6 cu.m	6.4 cu.m.
50	4.0 cu.m	16.00 cu.m.
100	8.0 cu.m	32.00 cu.m.
500	40.0 cu.m	160.00 cu.m.
3000	240.0 cu.m	960.00 cu.m.

Source: CGWB.

Rain Water Harvesting through Percolation Tank

Percolation tanks are constructed preferably on second to third order streams, located on higher fractured and weathered rocks which have lateral continuity down stream. Percolation is an artificially created surface water body, submerging in its reservoir a highly permeable land so that run-off is made to percolate and re-charge the ground water storage.

The re-charged area down stream should have number of wells and cultivable land to benefit from augmented ground water.

Recharging of Dug-wells

Existing and abandoned dug wells may be recharged after cleaning and desilting. The recharge water may be piped from desilting chamber to the bottom of the well or below the water level to avoid scouring of bottom and entrapment of air bubbles in the aquifer. The recharge water is made to pass through a desilting chamber or the filter chamber. Periodic chlorination should be done for controlling the bacteriological contamination.

Rain Water Harvesting through Check Dams/Cement Plugs/Nala Bunds

Check dams are constructed across small streams having gentle slope. The selected site should have sufficient thickness of permeable bed or weathered formation to facilitate recharge of stored water within short span of time. The height of check dams is normally less than 2m. and excess water is allowed to overflow so as to avoid scouring from excess run-off; water cushions are provided at down stream side.

Sand/Clay filled cement bags are arranged as a wall to check water across small nallahs. Some times shallow trenches are excavated across the nallah and ferro cement/asbestos sheets are placed on both sides of bags and gaps between bags and sheets are filled with clay.

Chapter 9

Biofertilisers

Fertilisers are applied to soil to replenish nutrients like Nitrogen (N), Potash (K) and Phosphorous (P) which is depleted from the soil by each harvest. Chemical fertilizers leave soils degraded, pollute and make it less productive. They also pose severe health hazards. Attempts to increase N-supply by excessive application of N supply lead to passage of N–to water bodies, nitrates to ground water and green house gases to the atmosphere. Consumer preferences have shifted towards organic foods grown without the use of chemicals. This awareness is not only growing in developed countries but also in India and other neighbouring countries. In India about 56.29 million hectares of land is either wasteland or fallow land which is proposed to be brought under cultivation. Thus total area under cultivation will be 170.29 million hectares. There is assured demand for biofertilisers to make agriculture profitable.

Biofertilisers include nitrogen fixers (symbiotic or non symbiotic bacteria), phosphate solubiling fungi and bacteria and mycorrhizal fungi, rhizobactaria that transfer the nutrients from soil to the plant roots. Biofertilisers like *Rhizobium Azotobacter, Azospirilium* and Phosphate solubilising bacteria fungi are used. Carrier materials such as peat, lignite, peat soil, wood charcoal favour the growth of micro organisms. Microbial inoculants like Azolla, Trichoderma, Frankia and Vesicular Arbuscular Mycorrhiza (VAM) are available in the market.

Micro-organism	*Nutrient fixed*
Algae	25 kg/N/ha
Azolla	900 N/ha
Azopirilium	10-20 kg N/ha
Rhizobium	50-300 kg N/ha
Azotobacter	upto 20kg N/ha
Mycorrhiza	Solubilise food phosphorous (60%)

Government of India provides non-recurring grant in aid of upto Rs.20 lakh for setting up biofertiliser production units of 150 MT annual capacity by entrepreneurs, NGOs, industry etc.

In India, present demand of phosphatic fertilisers is 47.98 lakh tonnes and chemical industry is able to supply 33.48 lakh tonne and balance of 14.5 lakh tonnes is imported. Mycrorrhiza provides phosphorus nutrition to the plants. These fungi utilises phosphorus from extremely low concentration and could offset the high cost of phosphate fertilisers. It can atleast partially replace chemical fertilizers. For technology one can contact. The Energy Research Institute (TERI), India Habitat Centre, Lodhi Road, New Delhi – 110003. For other Biofertilisers one can contact IARI New Delhi and other Agricultural Universities.

Vermicompost

After independence India faced a problem of food scarcity due to its high population growth. The solution of this problem was one of the main issues before the planning commission. It was felt that this problem can be solved by increase in the agricultural production in the country. Agricultural production increased by the introduction of high yielding variety (HYV) seeds, fertilizers, pesticides, crop rotation, development of irrigation facilities and mechanization. Use of various inputs improved the agricultural production and productivity but simultaneously created environmental problems.

The use of fertilizers and pesticides proved more harmful to soil as well as crop itself. Intake of these chemicals gave rise to several types of side effect to human beings. These problems led the scientists and researchers to revise techniques, which would not be harmful to people as well as environment. They have developed various location and climate specific technologies to overcome the problems to some extent.

The development of organic farming is one of the important technologies in this direction. The vermi composting is one of the techniques of natural sustainable farming. These techniques use waste household organic materials of farmyard and agricultural fields as raw materials and produce environmental friendly manure for agricultural production. It generates alternative means of earning and it reduces cost of farm inputs.

The chemical analysis of vermi compost shows that it has high level of nitrogen, phosphorus and potassium, which are necessary for growth and development of plants. In the present scenario quantity of solid waste organic material increases day by day in rural and urban area. Earthworms convert all types of waste organic materials into useful manure. They are unable to digest rubber, plastic, metals, stone and glass etc. Due to this reason these materials should be removed from the waste materials before use for this purpose.

Improvement Over Traditional Technology/Process

The farmers are using dung as manure directly without their decomposition in proper manner in the region, which is not giving appropriate result and increases the population of weeds. In this method large proportion of cow dung also are washed away with surface water without improving the nutrition power of the soil because it is not getting appropriate time to decompose. Sloppy lands constitute major problem of run off.

Some farmers used to put their cow dung and household waste materials in the pit. But due to lack of moisture for a long period it did not properly dissociate and not fully convert into useful manure. All these problems emphasize the need to introduce vermi composting techniques for high agriculture production in the region.

Process

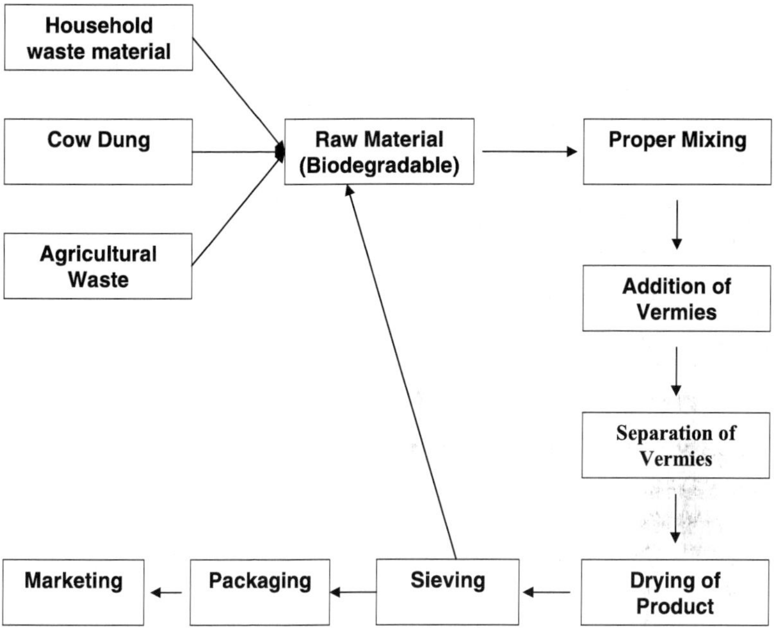

Time Schedule

Period	*Pit*	*Process*
0-30 days	pit-1	collection of biomass and cattle dung
30-60 days	pit-1	soaking of biomass with water, cowdung slurry and covering with black polythene sheet.
	pit-2	collection of biomass
60-90 days	pit-1	inoculation of earthworms
	pit-2	Biodung preparation
	pit-3	Biomass collection
90-120 days	pit-1	vermicompost ready; migration of earthworms to pit 2
	pit-2	vermicomposting
	pit-3	biodung preparation
	pit-4	collection of biomass
120-140 days	pit-1	harvesting of compost and collection of biomass
	pit-2	vermicompost preparation and migration of earthworms into pit-3
	pit-3	vermicomposting
	pit-4	biodung preparation

Techno economics

(a) Infrastructure:

(1)	Cost of shed (35x25ft 875) Sq.ft. =		Rs. 5,000
(2)	Cost of water tank 2000 Lit. capacity =		Rs. 3,000
(3)	Equipments:=		Rs. 5,700
	Tasla, Phavda etc.	Rs. 500	
	Sewing Machine	Rs. 3,000	
	Sealing Machine	Rs. 2,000	
	Cost of two steel buckets	Rs. 200	
(4)	Preparation of Beds (10 beds)		Rs.1,500
(5)	Cost of Worms(1500xRs.0.5x10		Rs.7,500
	Total		**Rs.22,700**

(b) Cost of Production (one cycle):

Cost of Cow Dung(100kgx10xRs.0.5) 1000 kg	Rs. 500
Cost of Waste material (200x10xRs.0.25)	Rs. 500
Bed filling and Removing compost labour charge	Rs. 500
Maintenance Labour Cost(45daysx1/2hr) 22.5hr/8	Rs. 150
Packaging Labour Charge (4xRs. 50)	Rs. 200
Cost of Packaging Material	Rs.2500
Total	**Rs.4350**
Total of Nine cycles (Rs.4350x9)	**Rs. 39150**

(c) Output:

(i)	Total Production of vermi compost(18kgx10) = 180kgx9 cycles = 16,200 kg @Rs.3/-	Rs. 48,600
(ii)	Selling of Earth Worms (60,000xRs.0.50)	Rs. 15,000
	Total	**Rs. 63,600**

(d) Profit:

Profit will vary in subsequent years

Ist year Rs. 63,600–(Rs. 22,700+Rs. 39,150)	Rs. 1,750
IInd year Rs.63,600-Rs. 39,150	Rs. 24,450
IIIrd year Rs.63,600-Rs. 39,150	Rs. 24,450
Total	**Rs.50,650**

The above profit is calculated on the basis of 10 beds. If number of beds increases the percentage of profit will improve due to scale of economy.

Source of Technology: Centre of Science for Villages (CSV) Sewa Gram, Wardha, Maharastra, M.P. Vigyan Sabha Bhopal and many other.

Utilisation of Fly Ash for Wasteland Development and Agriculture

In India the thermal power plants release nearly 118 million tone of fly ash for generation of power which has become major environmental hazard. Fly ash has been found useful for various applications and one such option is use of fly ash in waste land development. AMPRI Bhopal has developed a technology package for bulk utilization of fly ash for wasteland development and agriculture. The effect of fly ash on soil depends on various factors like reactivity of ash and soil, morphology–texture and structure, mineralogy, pH, ion exchange capacity, method of application, moisture content etc.

Fly ash helps in modifying the soil characteristics like decreasing the bulk density and increasing the water holding capacity, thereby saving irrigation water requirement. The major nutrients like phosphorus, potassium, sulphur, calcium and magnesium and micronutrients like copper, zinc, manganese, iron, molybdenum present in Fly ash have contributed to increase the fertility of soil. This soil modification and supply of nutrients converts the wasteland into agriculturally productive land and increases the productivity of various crops, cereals and vegetables.

The edible yields of tomato, cabbage, potato, wheat, pea and maize were more in ash mixed plots as compared to the control plot. Morphologically also the growth of plants in the pond ash admixed soil was better than that of control plots.

The impact and safety aspect of application of Fly ash in agriculture has been evaluated by testing the soils, food produces and ground water for trace and heavy metals like cadmium, chromium, cobalt, nickel, lead, arsenic, selenium etc. The radio-active parameters like actinium, radium, potassium have been analysed at Institute of Physics, Bhubaneshwar. No toxicity has been observed in soil as well as food crop produce. The agriculture produce grown on Fly ash amended soils meet the food quality standards prescribed by National Institute of Nutrition (ICMR), Hyderabad.

AMPRI has also organised farmers awareness programmes, Kisan melas at different part of the country and generated confidence among farmers and common masses on use of fly ash in agriculture.

Source of Technology: Advance Materials and Processes Research Institute (Earlier RRL) Bhopal.

Chapter 10

Tissue Culture

Tissue culture technology covers a wide range of techniques including *in vitro* culture of organs (shoot tips, root tips, flowers, ovaries, ovules, embryos, anthers etc.) tissue cells and protoplasts. Tissue culture has emerged as a valuable tool for mass propagation of several economically important species of medicinal and aromatic plants, spices, flowers, fruits etc. for generating disease-free stocks, stress tolerant varieties of higher yielding crops and for long term storage of germ plasm. The plant cell tissue culture has generated immense impetus to biotechnological research.

The beginning of tissue culture dates back to 1920's when German Botanist "Haberlandt initiated culture of leaf cells in Knop's solution. Subsequently, the modern scientific investigations advanced and improved the traditional skills required to propagate *in vitro* and continuing with artificial seed production.

Tissue culture has proved to be a good technique for mass propagation of a number of plants covering horticultural crops, floriculture, forestry species medicinal and aromatic plants which are rather difficult to propagate by conventional methods. Tissue culture was considered to be complicated technology but with passage of time several RandD Institutions and voluntary organisations have made efforts to simplify the procedures and have avoided the use of costly and sophisticated equipment so that educated farmers could adopt the technology at field level or rural level by taking 3-4 month training in organisations like NCL Pune, CIMAP, Lucknow, NBRI, Lucknow, TERI Gurgaon, Institute of Himalayan Bioresource Technology Palampur, COSTFORD, Kerala, ARTI, Pune, IIHR Bangalore etc.

There is no universal technology for the propagation of all plant species. Thus, separate protocols have to be developed for every species. The developed protocols are tested for reproducibility and transferred to persons for propagation. Multiplication is a purely routine activity which can be carried out by any young person having some basic educational background. The basic difference in propagation of different species is due to difference in the chemical composition of culture medium used at various stages of growth. A person having acquired basic training in mass propagation techniques can propagate any species after familiarisation with that particular protocol.

Following steps are involved in micro-propagation through tissue culture techniques:

1. Preparation of the appropriate sterile culture medium.
2. Culturing of the selected plant parts after surface sterilisation.
3. Manipulation of chemical composition of the medium for modulation of stage of growth.
4. Development of plants/shoots.
5. Rooting of the plant
6. Hardening to withstand external environment.

Cost of Equipment for a moderate Tissue Culture laboratory is around Rs.5.00–Rs.6.00 lakhs. Equipment like laminar air flow, air shower curtain, autoclave, inoculation cabinet, rotary shakers, low speed centrifuge, pH meters, balance, microscope, oven, glass ware, UV lamps etc are required. All the equipment is easily available from companies dealing in scientific instruments companies. Micro nutrients, organic and inorganic chemicals required in growth of tissues and seedlings are also easily available from dealers of chemicals.

All these costly equipment have been replaced by simple tools and devices in establishing rural tissue culture labs. Hood with UV Lamp, replaces laminar air flow unit. High humidity chamber can be made locally by fixing humidifer. Pressure cooker and stove replaces autoclave used in conventional process, temperature and humidity control are achieved in green house for hardening the plants which can be done in make shift polyhouses set up in the shade of trees. Other items include balance, measuring vessels, glassware and miscellaneous items like knife, forceps, spirit lamp, cotton, hand sprayers etc. As regards chemicals LR grade chemicals are replaced by commercial grade chemicals once protocol is developed and repeatability is achieved. A small laboratory could be established at a low cost of Rs. 25,000-30,000.

Low-Cost Tissue Culture Lab: Success Story of IHBT, Palampur

Department of Biotechnology, Government of India sponsored a project to be implemented by the Institute for undertaking the challenging task of capacity building of rural women in the modern techniques of plant tissue culture which was hitherto considered a very sophisticated technology. The Institute selected two groups comprising 25 women each hailing from two different locations in the Kangra district of Himachal Pradesh.

In the first phase these groups were familiarised with the activities of IHBT in the field of Tissue Culture and they were educated about the advantage of plants raised through Tissue Culture. The whole technology was published in the form of manual in Hindi covering text and illustrations. Subsequently, a mobile laboratory fitted with a custom made laminar air flow with a backup generator set was pressed into service and scientists visited these two villages periodically to train women in the process of Tissue Culture starting from media preparation, auto claving, explantation to inoculation under aseptic conditions. Realising the need of reducing contamination because of movement of culture to their homes which were 25-25 km away from place of work two separate units complete with laminars, culture racks and mini-auto-claves were setup. This enhanced the success rate of sterile cultures. The women groups hired rooms in the villages and shared the rent, electricity bill and water charges. A small hardening facility in the form of poly house was established for stabilisation of plantlets raised through Tissue culture.

These women groups are now proficient enough to work independently and concentrating on high value floriculture. Such efforts can be replicated elsewhere in the country and tissue culture will no longer remain within the four walls of high tech laboratory or industry.

Chapter 11

Low Cost Nursery

Introduction

Production of ready to transplant seedlings of vegetable, fruit, ornamental, forestry and agricultural plants/crops (if possible) in nursery and make them available for transplantation in suitable and ready field at right time is a good income generation activity for the rural/tribal women and men. The scope of this activity is widening because of increasing demand of high quality planting material. Seedlings may be prepared by two methods:

1. In plastic bags, and
2. In sand beds.

Plastic Bags

In the case of woody dicots like *Jatropha, Salvadora*, cotton, arhar, castor and fruit, ornamental and medicinal plants and sugarcane the seedlings should be produced in plastic bags.

Sand Beds

In case of most of the vegetables and cereals, the seedlings can be produced on a bed of sand. By this method, possibility of growing weeds is very less because of non mixing of compost and soil in the sand bed.

Vegetative Propagation Through Rooted Cuttings

Root cuttings of most tropical dicot species have the capacity to produce roots if they are properly treated. Such rooted cuttings have the same genetic constitution, and therefore, the same characteristics, as the mother plant. Cuttings of most vines, lianas and herbs, and also those taken from young saplings of trees, root more easily than cuttings from adult trees (*i.e.*, trees that are already producing flowers/fuitis). In general, soft wood cuttings (*i.e.* those having green bark) root more easily than hardwood cuttings (cuttings having brown bark). Spring time is the best season for making rooted cuttings, because the cells are in active state of growth during this season. If the cuttings are to be transported over some distance or if it is not possible to treat them immediately, these should be kept wrapped up in wet gunny cloth. Before planting the cuttings, they should be cut into sections, about 10 to 12 cm. long. Leaves on the upper most node should be left on the cutting while those on the lower node are to be removed. If the leaves are very large, the laminae may be cut into half to reduce their size, which allows more cuttings to be allowed in the same area. Hormonal treatment of the basal part of the cutting increases the percentage of rooting. The treatment consists of dipping the basal (1 or 2 cm. of the cutting for about 10 sec. in 4000 PPM solution of indole butyric acid (IBA) or naphthalene acetic acid (NAA) and planting such cuttings in moist sand. For preparing 4000 PPM IBA solution, first prepare 5 per cent KOH solution, weight 100 mg IBA and place in it a beaker. Add the KOH solution drop by drop till the IBA dissolves completely. Add sufficient water to make up the total volume to 25ml. NAA is commercially available as a 4.5 per cent liquid formulation. By diluting it with 10 times the volume of water, one gets a 4000 PPM solution.

For good root development, it is necessary for the basal part of the cuttings to be supplied with adequate moisture and oxygen. This is achieved by planting them on beds made of fine sand. Before preparing the bed, lay a black plastic film (1.5 M wide and having any convenient length) on the ground, place a single layer of bricks along the border of the plastic and fill the intervening space with sand. If cuttings are slanted into this bed they would produce roots faster than in beds made of soil. Water the beds daily till excess water starts draining out of the beds. It is essential to prepare the beds under the shade of a large tree or underneath a shade netting.

High Humidity Chamber

Cuttings planted for rooting can lose moisture if the weather is dry. This can be avoided by keeping the cuttings in all atmospheres having more than 70 per cent relative humidity. If the rooting experiments are conducted in the coastal regions, the atmospheric humidity is generally very high. But in inland regions, where the atmospheric humidity can go below 70 per cent, the beds, into which the cuttings are planted, are covered by a transparent plastic film, supported by wire arches. The atmosphere within the space covered by the plastic film is very soon saturated by water vapour, indicated by droplets of water, condensed on the inner surface of the plastic film. Remove the transparent plastic cover once or twice in a week and water the beds till the excess water starts draining out of the beds. Twigs planted in this chamber remain turgid and fresh even upto 20 weeks. The percentage of rooting is dramatically increased by using a chamber of this type. The only precaution that one has to take while operating this kind of a high humidity device, is to place it in a shady place. If the high humidity chamber gets exposed to direct sunlight, it would get so heated up, that the twigs would die of thermal shock.

The high humidity chamber can also be used for making grafts of various kinds. The scion remains green and turgid for a longer period in the high humidity chamber and therefore the survival percentage of the grafts is increased.

Hardening of Rooted Cuttings and Grafts

The cuttings that have sprung roots, are uprooted from the sand bed by inserting four fingers of the hand below the cutting and scooping out the cutting along with the sand. The hand holding cutting is then dipped into a bucket full of water and moved gently to and from to wash the sand off the roots. The cuttings is then placed into a plastic bag or a small earthenware pot filled to about 3/4th of its height with potting medium. The roots should rest on the potting medium. The rest of the bag/pot is then filled to capacity with potting medium and the bag/pot is well watered. Such a bag/pot is then kept in the high humidity chamber for about a week, so that the plant recovers from the transplanting shock. After this period, the plastic cover of the high humidity chamber is lifted gradually in such a way, that after about 8 days, the plant is exposed completely to external dry air. This process is called hardening and it is necessary before one can expose the sapling to external atmosphere.

Low Cost Greenhouse

With the help of transparent polythene sheet and treated bamboo (sodium dichromate + boric acid + copper sulphate) low cost green house can be prepared which helps in accumulation of over night carbon-dioxide resulting in higher rate of photosynthesis and ultimately better growth of plants.

Scales of Operation

Capacity

Capacity of Nodal Centre is 24000 plants of different species including vegetable plants, fruit plants, medicinal plants and ornamental plants etc. Size of the land is 100 sqm., out of which 10 sqm. is for walkways, 30 sqm. for green house and 60 sqm. for plants. Capital investment will be over Rs 20,000-25,000 and profit is Rs. 20,000/yr.

- ☆ *Manpower Requirement*: Manpower required for one cycle is 48 man-days worth Rs. 2104.00 which was contributed by members themselves.
- ☆ *Packaging*: Vegetable plants are directly distributed to the farmers from nursery bed according to their demand while plants of rainy season and fruit plants are distributed to the farmers in polythene bags of 3"x5", 5"x7", 4"x6" sizes as per plant variety.
- ☆ *Other requirement*: Fencing is very essential at nursery site to protect the plant from stray cows.
- ☆ Working Capital/Profitability:

The nursery business in highly remunerative and can be conducted in small space with minimum investment, making it highly suitable for adoption by small/marginal farmers. In nursery of this kind area of 1 sqm. accommodates 400 seedlings in polythene bags of 3"x5" and provides an average gross annual income of Rs. 2200.00 per sqm. This business can be easily taken up in spare time, using very little space and inputs.

The nature of the nursery can vary according to the needs of the locality. A nursery in the vicinity of urban areas may produce mainly ornamental plants, indoor/garden plants, avenue trees and trees with showy flowers, while in rural areas a nursery may make more money through supply of grafts of fruits trees, rooted cuttings and saplings of medicinal and aromatic plants.

Source of Technology: Appropriate Rural Technology Institute (ARTI), Phaltan, District, Satara (Maharastra), is the source of technology.

Chapter 12

Employment Generating Technologies

Detergent Powder and Cake

Detergent powder and cake has replaced washing soap in rural and urban homes as it has better cleaning properties. Typical formulation for popular grade detergent is as follows:

Acid Slurry	10 Kg
Soda ash	56 kg.
Sodium sulphate	20 kg
Colour	300 gm
Optical whitener	200 gm
Perfume	150 gm
Water	13 litres.
	100 kg.

Process for Detergent Powder

- Mix acid slurry, water and colour.
- Add soda ash slowly with thorough mixing.
- Add sodium sulphate
- Finally add optical whitener, perfume.
- Pack in 500gm/1kg LDPE bags.
- PVC Tubs or concrete tanks and wooden stirrer could be used.

Process for Detergent Cake

- Mix acid slurry, water and colour.
- Add soda ash, talc in kneader
- Mix with china clay, starch, sodium silicate and STPP.
- Finally add melted wax, perfume, and optical whitener and mix well.
- Total kneading time is 40 minutes.
- Extrude the mix, cut into cakes of 125/200gm. and wrap in the coated paper.

Process for Liquid Detergent (100 L)

Acid Slurry	15 kg.
Caustic lye (50 per cent)	5 kg
Urea	5 kg
Colour	200 gm
Perfume	200 gm.
Water	75 L

Process

- Mix Acid slurry and water.
- Neutralise with caustic lye. to pH 6.5-7.0
- Add colour, urea and water to make 100L.
- Pack in polytheylene Jars.

Note: All the ingredients are available from Ahmedabad at competitive rates extruder, kneader, mixer are available from M/s Chemestar Pvt. Ltd. Ahmedabad and several other vendors.

At rural level the mix could be pressed using locally fabricated press and sun dried.

Carcass Utilisation

Indian possesses the highest livestock population in the world. However, the problem of disposal and utilisation of carcass of dead animals has not received adequate attention in rural areas. As a result the country is losing Rs. 1000 crores annually due to non-recovery of the useful waste materials like raw hides, meat meal, bone dust, tallow etc. from the fallen carcasses of livestock and other animals. Improper recovery of flesh, bones and intestines lead to the loss of useful inputs for the production of meat meal and bone dust and tallow. It is essential to provide technological inputs to traditional leather workers.

Charmkar is the caste that works with cattle skin performing various operations from flaying to making finished products out of processed skins. However, on account of gradual deterioration of the sector, the Charmakars in the rural areas has been reduced to a state of penury, pushing them below the poverty line and in many cases leading to their abandoning the traditional trade and making up for agricultural labour in the neighbouring areas.

Improvement Over Traditional Technology/Process

At present carcasses of the fallen cattle in rural areas are usually taken to the designated place outside the village and after flaying left on the spot as carrion except bones which constitutes 20-25 per cent of body weight. They are collected by traders for supply to bone mills and hides are supplied to tanners while residual constituents that cause environmental and health hazards remain to decay.

A technology package has been developed by CLRI, Chennai for utilization of entire fallen carcasses for following uses:

- ✰ Meat meal: Poultry feed supplement and fisheries
- ✰ Bone meal: Poultry feed supplement and for pet animals.
- ✰ Bone powder: DCP fertilizers, Gelatin
- ✰ Cracklings: Pig feed

- ☆ Horns, hooves, hair: Handicrafts
- ☆ Bone ash: Bone-china POTTERY
- ☆ Fine bone ash: Medicinal purpose
- ☆ Raw hides: Tanning industries

The complete processing of the fallen carcass is shown here in the flow chart:

Flow Chart

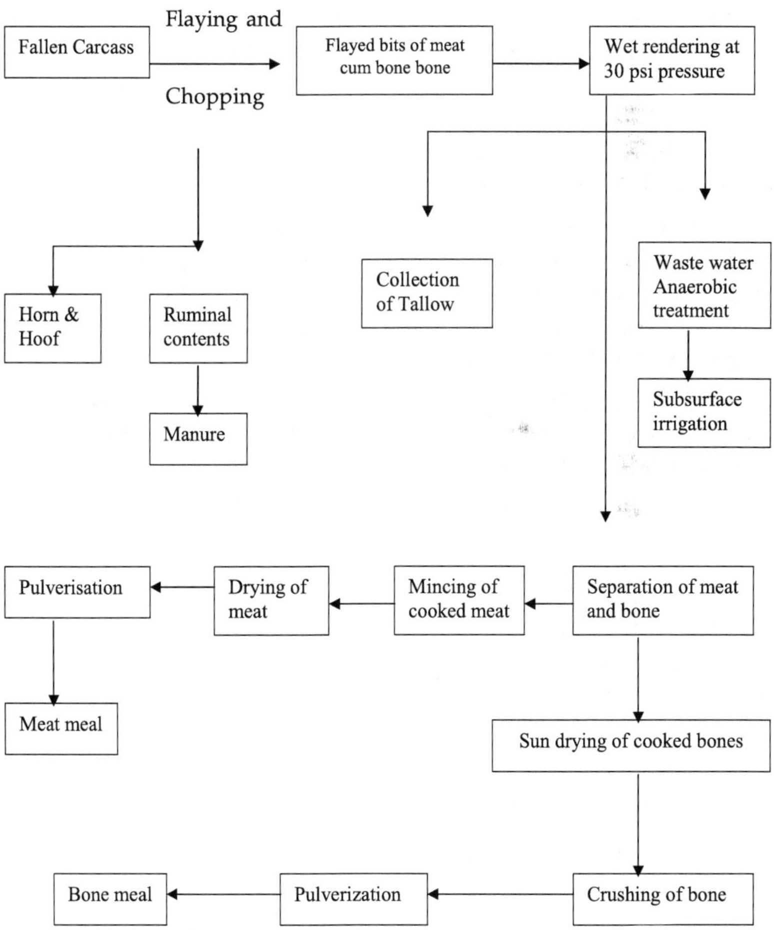

Total carcass utilization through a series of mechanical operations would enhance the income of socio-economically backward flayers, provide income to 2-3 persons, improve environmental conditions, generate revenue.

Process

The carcass is carefully lifted and brought to the carcass utilization unit using locally available horse/bullock cart covering the distance of 5-6 kms. Low cost motorized carriers are also used for this purpose.

The carcass is then stripped of visible fatty tissues, which are separately heated, clarified and strained to yield 50 per cent by weight of pure, grade-A fat/tallow; creamy in consistency and white in colour. The cracklings are used as pig feed. Offal are removed and put in compost pits/tanks along with blood and to residual matter to generate manure and ensure clean environment. The rest of the carcass with bones is chopped and put in cooker and cooked under pressures of 35 p.s.i. for about 2 hr. Fat/Tallow of B grade yellowish or brown in colour is obtained at the base of the cookers. After the cooker is cooled, cooked flesh and digested bones are removed and easily separated.

The flesh is dried for removal of 80 per cent moisture using indirect heat obtained from the cookers. The cooked meat is minced before drying using motorized mincers. This gives dry powdery meat meal free from infestation even after extended storage. The digested bones are sun dried, mechanically crushed in a crusher and sieved to fine grain as bone meal.

Capital Investment:

	Capacity	3-4 carcasses/day (in lakhs)
A.	Land ½ acre	0.50
	Work shed/civil work Stores	3.50
B.	Equipment	5.50

Working capital (Annual)

No. of working days–300

Capacity–3-4 carcasses/day. (in lakhs)

1. Raw material @ Rs.200/carcass x 300 x 3 = 1.80
2. Manpower

Manager–1	4000/pm	0.48
Skilled workers-2	2000/pm	0.48
Field workers–2	1500/pm	0.36
Watchman-1	2000/pm	0.24
		1.56
Fuel (Electricity)		1.00
Maintenance		0.50
		4.80
Depreciation, overheads, interest etc.		2.04
		6.90

Return on Investment: (in lakhs)

Sales	10.50
Prod. cost	6.90
Profit	3.60

$$\% \text{ Profit } \frac{3.60 \times 100}{9.50} = 38 \text{ per cent}$$

Source of Technology: CLRI, Chennai

Queen Pottery Kiln (Manufacture of Pottery ware)

Regional Research Laboratory, Bhubaneswar has developed a pottery kiln for firing green pottery ware, artifacts etc. in which zero defect efficiency has been achieved due to easy flow of flame. Temperature of 1000°C is achieved in 2-3 hrs. by using coal as fuel. Design of 4.5 cum and 5.5 cum. capacities are available. It is possible to fire pottery ware, terrakota artifacts, handicrafts, water filters, flooring and roofing tiles. For firing items made from 100 kg. of clay, 250 kg. coal is required and it takes 7 hours which is 75 per cent less than kilns using biomass saving 70 per cent fuel. 10,000-15,000 bricks are required to construct the kiln. Raw materials required are red bricks, clay, sand, iron rods, wire mesh etc.

Potters can also adopt improved potter's wheel available from State KVIB/KVIC fitted with wall bearings or single phase electric motors: central Glass and Ceramic Research Institute, Kolkata has developed pedal operated potters wheel offering ease of operation.

Techno-Economics

Capacity	100 firings per year (10,000 items)
Land and building	on rental or ownership basis
Cost of Kiln (2.5cum)	25,000
Potters wheel	7,500
	32,500
Working Capital	15,000
	47,500
Say Rs.	50,000
Sales	1,00,000
Profit	50,000

Cattle Feed/Poultry Feed

There is growing demand for cattle/poultry feed in rural areas due to efforts of Govt. to enhance milk and poultry production under various employment generating programmes. There is need for balanced cattle feed in addition to green fodder and hay for general improvement of health of cattle. Availability of well balanced feed containing carbo-hydrates, protein, vitamins and minerals has become necessary. Keeping in view the locally available raw materials a balanced formulation can be developed for production of cattle feed.

The process of manufacture of cattle feed and poultry feed is relatively simple and consists of reduction of size through pulveriser and blending of various ingredients according to formulations. Ribbon blender is used for uniform mixing. The vitamins/minerals are added during mixing. The mix is passed through an extruder to get pellets which are finally packed in HDPE/gunny bags. Entrepreneurs should take help of Veterinary Department for formulation to maximise use of local raw materials. A typical formulation is given below:

1.	Ground Nut/Soya bean cake	30 per cent
2.	Wheat bran or pulse husk	10 per cent
3.	Rice bran or cotton seed bran	25 per cent
4.	Maize	10 per cent
5.	Damaged wheat	10 per cent
6.	Cotton seed	2 per cent
7.	Molasses	10 per cent
8.	Salt	1 per cent
9.	Calcium carbonate	0.5 per cent
10.	Mineral mix.	1.5 per cent
11.	Vitamin mix	0.1 per cent

Capacity 5T/day

Land/Building

Land 10,000 sq.ft.	1.00 lakh
Building 2000 Sq.ft.	4.00 lakh

Machinery/Equipment

1. Disintegrator/Pulveriser–500 kg/hr 2. Ribbon Blender–500 kg/hr. 3. Gravity sifter–500 kg/hr. 4. Weighing platform 5. Gunny bag sealing machine	5.00 lakh
Utilities/Electricity	1.00 lakh
	11.00 lakh

Entrepreneurs can also contact CFTRI, Mysore for technology.

Neem Phenyl: A Herbal Product

Neem is widely available throughout the country as wild growth or planted in villages/suburbs for shade. The pesticidal and insecticidal properties of Azadiricachitin present in Neem have been internationally acclaimed now although Indian Homes using neem leaves from time immemorial for preservation of food grains and woollen clothes.

There is great stress on use of herbal products in the country as well as all over the world due to absence of any side effects.

It is suggested to manufacture Neem Phenyl for use in houses, hospitals etc. in place of chemical phenyl. The product would find wide market. The ingredients used in this formulation would also repel flies, mosquitos and cockcroaches.

Raw Material

Following raw materials are required in the manufacture of Herbal Phenyle which are easily available in the local market.

	Rate
1. Neem Oil	Rs. 60/per litre
2. Water	—
3. Citronella Oil	Rs. 400/per litre
4. Colour	Rs. 700/per litre

Material Mix for 100 Litre of Phenyle

Material	Percentage	Quantity	Cost
1. Water	90 per cent	90 litre	Rs. 2.00
2. Neem Oil	5 per cent	5 litre	Rs.300.00
3. Citronella Oil	1 per cent	1 litre	Rs.400.00
4. Colour		200ml	Rs.140.00
5. Packaging materials			Rs.1000.00
		Total:	Rs.1842.00

Say Rs. 1900/- for 100 litres

Machinery Required

1. Mixer-cum Emulsifier of 200 litre capacity. — Rs.25,000
2. Packaging machine — Rs. 10,000
3. Measuring cups and misc. — Rs. 5,000

Process of Manufacturing

Raw materials–water, Neem oil, and citronella oil are mixed in proper proportions and transferred to emulsifier. The emulsification process is carried out for 4-6 hr. The emulsified material is filled in PVC bottles/cans and sealed by using cap sealing machine.

Indigo Blue-Dye

The use of natural colours for human use exists from time immemorial. In different regions of our country people follow their traditional methods for preparing natural dyes. The realisation of hazards caused by synthetic dyes has made the society to revert back to traditional raw materials.

MCRC has been involved in extracting dyes from different natural sources for more than a decade. The dye pigments from bacteria, indigo blue dye from *Indigofera tinctoria, Writtia tinctoria,* and *Tephrosia* sp., red and orange dyes from *Morinda tinctoria* and *Bixa orellana* respectively and brown dye from *Acacia catechu* are the major scientific achievements of the Centre.

For blue natural dye the only viable choice is indigo from plants. The plant is grown in Asia, Africa, the East Indies, the Philippines and America. In MCRC process natural indigo is obtained by fermenting the foliage of species of Indigofera and a good tech-pack is now available for use. All the processes involved in this package of practice are ecofriendly and cost-effective.

The extraction of indigo from the plant consists of mainly 3 stages *viz.* steeping the plant in water (for fermentation), separation of the aqueous extract and oxidation of this solution with air and separating the precipitate and preparation of marketable dye cake or powder. MCRC has developed a new microbial process which has been found efficient for extraction of indigo dye. A patent has been filed on the technology package. Under CAPART's project more than 100 various Indian organisations have been trained on natural dyes, especially on Indigo.

MCRC's Indigo extraction technology package, consists of:

Organic Plant Production

Cultivation method for indigo plant includes crop maintenance and crop protection by biological control and use of residual green matter as green manure and for compost preparation.

Dye Extraction Techniques

Harvest of plants, microbial fermentation of plants, culturing selected bacteria by simple method, use of netlon soaking set-up, mechanical oxidation, slurry collection, boiling and filtering.

Technical Know-how

Construction and design of indigo tank (MCRC Model), fabrication and maintenance of indigo agitator and fabrication of netlon soaking set-up.

Aspects of Indigo Market

Indigo plant, dry leaf, indigo dye and residual matter are sold in the market. For MCRC's indigo dye extraction, initial investment of around Rs. 60,000/- for a unit is required. Minimum 15 acres of land are required to support an extraction unit for 200 working days in a year.

Economics

Cost of cultivation	Rs.5000/- per acre
Biomass obtained	4-6 tonnes
Return	5000 kg @ 2/- per kg.
	= Rs. 10,000.00 per acre
Processing cost	Rs. 1/- per kg.
Toal cost of Production Rs.(2+1)=3	Rs.15,000/-
Sale value received @ Rs.4/-per kg	Rs.20,000/-
or Rs. 400 per kg of indigo	
Sale of residual green matter	2,500/-
@ Rs. 0.50 x 5000 kg.	

Total sales	22,500/-
Profit	22,500–15,000
	= Rs. 7,500/- per acre

Profit per year from 15 acres 7500x15= Rs. 1,12,500/-

Cattle Lick Salt

Apart from general feed, milch cattle require minerals for their healthy growth, nourishment and increased yield of milk. Generally, salt is given to animals along with feed to save them from indigestion and stomachic problems. If the salt is supplemented with micro minerals like Ca, Mg, Zn, Cu it gives good results. Central Salt and Marine Chemicals Research Institute, Bhavnagar has developed mineral-salt blocks which animals lick for good health. These are being produced commercially for dairies, inviduals and wild life parks.

Process of manufacturing is very simple. Minerals in required ratio are mixed in salt and slurry is made by addition of water. Wooden mould of special design are filled with this mixture and hammered with wooden hammers for compaction. The moulds are then sun dried for 8-10 hours and ejected from moulds after becoming dry. Mould could be made by village carpenters and raw materials are also easily available. Formulation is available from CSMCRI Bhavnagar.

Capacity	200 blocks/day weight 2.5 kg each
1. Land land Building	Rental
2. Machinery/equipment and moulds	40,000
3. Working capital	1,00,000
Total investment	1,40,000

Japanese Quail Breeding

Quail is the most efficient machine for converting its feed into animal protein of high biological value having amazing taste, flavour, low calorific value and high dry matter content. It is also the cheapest animal protein for human diet suitable for pregnant women and children. Its meat enhances immunity against diseases like arthritis, diabetes, cancer, asthma, tuberculosis etc. It's eggs are one fifth of the size of a hen's egg. As such these are within the reach of economically poorer section of the society. It's rearing is very popular in countries like Japan, China, Singapore, Hongkong, France, England, Italy, Spain etc. It started in India just a decade or so back. Quails are highly resistant to poultry diseases and no vaccinations are required. Japanese variety is fast growing species and attains a weight of 150 gm. within 6-7 weeks. These quails attain sexual maturity within 42-45 days and gives 260-280 eggs per female quail. It also requires less space and feed compared to poultry birds. Quail meat and eggs is very nutritive and contains minerals, vitamins, fat and metallic ions.

The Japanese quail has been permitted for rearing vide G.O.I. letter 203-22184 FRY (WL) dated June 27, 1997 and powers are vested with officers of State Deptts. of Animal Husbandry (Veterinary Asstt. Surgeon) under Wild life protection Act of 1972. Farmers and unemployed youth in rural/semi urban areas should go for quail rearing which is less capital intensive and profitable.

Quails can be reared in cages of 60x60 cm in which 20-25 quails could be easily accommodated. Quails can also be reared in multitiered cages to economise on space. Temperature in the cage should be between 13-14° C and 37-38° C as production is affected and mortality is increased beyond 43.3° C. As such cooling/heating arrangements are to be provided. Lighting may be provided in the night to enhance productivity.

Economic Feasibility

Capacity–500 Quails/week or 2000 Quails per month.

A. Fixed capital:

(i)	Land and Building 500 sq.ft.	60,000
(ii)	Brooder/Cages	10,000
(iii)	Net and utensils	5,000
		Rs. 75,000

B.	Working capital	
(i)	2000 chicks @ Rs.2/- of one week	4,000
(ii)	Feed/month @ 500 gm/quintal @ Rs.8/- per kg	8,000
(iii)	Misc. and Utilities	1,500
		13,500

C.	Sales of 1800 Quails.	
	Assuming 10 per cent mortality @ Rs.15/per quail	27,000
	Profit per month	27,000–13,500
		Rs. 13,500/-

Farmers can contact local hatcheries or can also take up hatching which takes 17-18 days.

Source of Technology:

1. Indian Avian Research Institute (IVRI) Izzat Nagar-243122, Distt. Bareli U.P.
2. Regional Research Laboratory, Canal Road, Jammu-Tawi.

Rabbit Farming

Rabbit Farming is profitable enterprise in rural and tribal areas. It is now persued in Himachal Pradesh, Jammu and Kashmir and Hilly areas of South India. There are two varieties of rabbits adopted for culture *i.e.,* (i) Angora for wool and fur, (ii) Chinchilla for meat. Angora wool has a good demand in woollen industry and price of 1 kg. of rabbit wool is about Rs. 1200 as it is still being imported to meet the demand of the Indian woollen Industry. Rabbit meat is a delicacy and has zero cholestrol content and hence suitable for heart patients. Angora breed should be reared in low temperature areas as rabbit rearing can be carried out in the temperature range of 5-25° C; otherwise artificial cooling has to be provided. Chinchilla breed can be reared in temperature upto 35° C.

It is a fast multiplying animal:

$$2M + 10\,F \xrightarrow{\text{one year}} 200\ \text{Nos.}$$

Pregnancy period is 30-35 days and a female gives a litter comprising 7-8 off springs every three months. Rabbits are herbivorous and consume green feed, soft hay grass, kitchen waste and also solid pelletized feed (15 per cent of feed). Food requirement of the animals is 3.7-4.0 per cent of body weight and energy requirement is 1000-1200 cal/kg. of body weight. Therefore, 100-200 gm of pelletized feed should be given daily alongwith green feed. For commercial scale breeding rabbits are kept in cages of 24" x 18" kept on 2-3 Tier racks. Sunlight and air is essential. Ideal temperature of the place should be in the range of 5-25°C. The cages should be away from populated area, and protected from noise, smoke, fumes and protected also from predators like dogs, cats, and jackals.

Health Management

Diseases like scabies, ear mange, weepy eyes, bronchitis and diarrhoea are common among rabbits. Good sanitation and proper feeding can minimise the diseases. Common medicines for human beings are given to rabbits restricting the dosage keeping the weight in view.

Rabbit is a very delicate species which should be handled gently avoiding jerks. Cleaning of water and feed bowls should be done regularly. Sufficient water also should be provided.

Mating should be done during early hours of morning and female should be brought to the male. After 15 days of mating belly of the female should be checked by gentle pressing. After 25 days bed should be prepared for pregnant doe. Separate the young from the mother when they start feeding on their own as some mothers eat their younglings. Special care should be taken for Angora rabbits by keeping them in separate cages in cool climate. In summer coolers should be provided.

Shearing of wool (2-3" long) should be avoided prior to winter months but completed before onset of summer. Normally 4 cuttings weighing 700-750 gms are obtained in a year per rabbit.

Techno-Economics (150 Nos. Keeping 10% Mortality)

Cost of (10 females + 2 males) rabbits @ Rs. 500/- per rabbit.	Rs. 6,000
Cost of Cages 150 cages @ 150 per cage	Rs. 22,500
Feed (in one year)	Rs. 30,000
Medicines and Labour (in one year)	Rs. 15,000
Total	Rs. 73,500
Say	Rs. 75,000
Cost of production	Rs. 45,000
Interest	Rs. 10,000
Sales (100 kg of wool @ Rs.1200/kg.	Rs. 1,20,000
Profit	Rs. 65,000
Or	Rs. 5,000 per month

Source of Technology/Training:

1. Sheep and Wool Research Institute, Gadsa, Dist. Kullu, H.P.
2. Regional Research Laboratory, Jammu Tawi, J and K.

Poultry Farming

Poultry farming is a viable business activity and has got immense scope for growth in India. It has greatly helped in the economic upliftment and employment of our rural population. Today, India ranks as the sixth largest egg producing country in the world. During the last three decades, annual output of eggs has gone up by over eight times, making poultry the fastest growing sector of Indian agricultural economy.

Although, India is the world's sixth largest egg producer, its per capita consumption is poor (less than 30 eggs per year). This situation is worse still for chicken meat (about half a kilogram). This potential for increasing egg consumption is enormous particularly in rural/tribal areas where about 75 percent of the population lives. The poultry products cost more in rural areas, as they have to be transported from urban areas. Almost 75 per cent of eggs and chicken meat is consumed in urban and semi-urban areas which account for about 25 per cent of the country's population.

Both eggs and chicken meat are rich in protein. Considering the high value of poultry products it is recognized as an important income generating source for the poor marginalized farmers and landless labourers. Poultry farming offers a great scope and has a large potential for economic development of rural masses.

The scientific methods of poultry farming are given below:

1. Preparation of site
2. Chick guard construction and lighting
3. Rice Bhusi with lime for flooring
4. Feeding apparatus and Water Pots
5. Medicines
6. Marketing

Process Flow-sheet

Hatchery

↓

Supply of old chicks

↓

Keep in baskets, provide feed and provide heat through Bulb or lantern

↓

Vaccination by local Vet. Doctor

↓ Expected 30 per cent mortality

After 12-14 weeks Broilers of 1 kg weight ready for sale

Economics

Capacity: 50 Boiler + 25 coiler

Power: Single phase electric connection or petromax

Fixed Cost

i) Chick guard, feed and water pots Bucket, Garana, weighing machine etc.	3,000
ii) Cost of Chicks	3,000
	6,000
iii) Working capital	8,000
(i)+(ii)+(iii)	14,000
Profitability	40 per cent

Spirulina

Spirulina powder is prepared from blue-green algae grown in water bodies/tanks under hygienic conditions. Spirulina is a very rich source of protein (61 per cent) and multi-vitamins like B, C and E. About one gram of *Spirulina* powder is equal to 1 kg of fruits and vegetables in terms of nutrition content. In Egypt people used *Spirulina* as protein rich food almost 5000 years ago. It generated interest recently after 1950 due to its use as space food. It has become a craze for slimming apart from the following uses:

- ☆ Treating anaemia
- ☆ Treating diabetes
- ☆ Lowering of cholestrol
- ☆ Healing of wounds
- ☆ Food colourant
- ☆ Protein supplement etc.

The process involves inoculation of spirulina culture in pucca tanks having mechanised agitators to oxygenate the water. About 20-25 gm of *Spirulina* grows in 1.0 sq. meter surface area of water in a day. The blue-green algae is removed from water surface and allowed to dry before purification and production of powder by spray drying process.

For production capacity of 6 Tonne/year of *Spirulina* powder there is fixed capital investment of Rs.36.00 lakhs which includes construction of water tank of 2000 Sq.m. and equipment like filtering unit, spray drier, vacuum packaging unit and packaging material. It is suggested that farmers may construct tank of size 20m x20m and grow algae under hygienic conditions and supply the green algae to mother plant having refining and spray drying facilities. This will involve large number of farmers and reduce individual investment and enable the plant to achieve higher production capacity. The mother plant could supply powder to pharmaceutical industry for packaging in the form of capsules/sugar coated tablets.

Technology can be obtained from CFTRI Mysore or MCRC, Chennai through NRDC.

Village Bakery

Bakery is considered as a traditional activity in the country as it occupies an important place in daily dietary habits. With growing population, improving nutritional standards and the demand for fresh ready to eat foods the demand for bakery items has grown. Even the demand of products like atta biscuits, coconut cookies, bread, buns (for burgers) has increased.

All the raw materials like wheat flour, maida, suji, vanaspati oil, sugar and milk etc. are available in rural areas. The equipment like grinders, mixers, moulding machine, country type oven, bread slicers, balance and other miscellaneous vessels are available. It is also possible to improve the traditional bakeries by training and technological inputs from organisation like CFTRI Mysore. In construction of kilns, refractory bricks could be used. An insulating layer can be provided to reduce heat losses.

Techno-Economics

Capacity	200 loaves/200 kg biscuits/day
Building	50 Sq.m.
Equipment	Rs.2.50 lakh.

Working capital	Rs. 2.00 lakh.
Man power	4 persons
Power	3 KW
Water	5000 L
Fuel	50 kg of coal/Diesel.

Improved Jaggery (Gur) Making

Gur making is an age old agro-based rural industry. Almost 50 per cent sugar cane is utilised for production of 80 lakh tons of gur which apart from imparting sweetness to food, contains nutrients and minerals.

In the traditional process, the sugar cane is crushed by roller crushers and juice is transferred to open pans heated by underground 'Bhatti's' constructed by using common building bricks and clay mortor. Normally, bricks are not used on the floor of kiln. In the improved process three open pans (Kadies) are kept over the kiln in sequence for boiling of sugar cane juice. The dried bagasse is fed from one end of kiln while on the other end chimney is provided for release of smoke. At present, due to improper design of kiln and chimney more fuel is consumed and dense smoke is released which comes out of chimney.

Indian Institute of Petroleum (I.I.P.) Dehradun has developed improved design of kiln and chimney which has following advantages.

1. Saving in fuel (bagasse) by 10 per cent
2. Reduction of smoke/flue gases from chimney as it is environment friendly.
3. Use of round chimney of suitable height instead of square cross section.
4. Longer life of kiln by use of refractory bricks.
5. 20 per cent more production of gur from same quantity of Juice.
6. Improvement in quality of gur.
7. Fuel feeding system has been improved and air entry ports have been optimised.

Economics

Production capacity: 8-10 q/day.

Cost of shed 400 Sq.ft: 1.00 lakh.

Equipment

1. Furnace and Chimney
2. Electricity diesel engine
3. Power driven sugar cane crusher
4. 3 Boiling pans (MS 62" dia of thickness 14, 16, 22mm.)

} 1.5 lakh.

A large number of improved gur making units have been set-up on IIP design in UP and Uttarakhand. No doubt the cost of construction of improved design of kiln is more but it could be compensated in two seasons.

Interested Persons Can Contact

Indian Institute of Petroleum, P.O. Mohkanapur Dehra Dun, Uttarakhand for Technology or demonstration of process in working units.

Handmade Paper Making

Conventionally paper is made by utilizing wood. However, excessive usage of wood for pulp and paper making results in denudation of forests leading to environmental problems. To avoid such a situation alternative cellulosic sources (non-woody substrata) have to be resorted. These alternative substrata can be the agricultural wastes such as sugarcane bagasse or straw belonging to paddy, wheat, maize, barley, ragi, millets, etc. Silk cotton (from *Ceiba pentandra*) has also been used for pulp and papermaking. Here, the plant is undisturbed but only the silk cotton is harvested from the fallen pods, which is renewable. In addition to these alternate sources, making of hand made paper recycles the waste paper available in plenty.

Eco-Friendly Paper Making

Conventional method of paper making uses cellulose in the form of wood chips obtained by felling trees, which has increased global warming. Moreover, in this method, chemicals are used which release toxic substances causing environmental pollution (air, water and noise pollution) requiring treatment of water. Surveys state that for manufacturing of 1 ton of paper 277 *Eucalyptus* or 462 bamboo plants are required. Moreover 1,00,000 greeting cards made with hand made paper can save 500 trees.

Equipment

1. Hand auto vat (Teak wood frame)
2. Aluminium handle 4" x 2" x 15"
    - Brass hook 5" x 2" x 15"
    - Rubber beading
    - Nylon mesh
    - Mixer
3. Hand made vat
4. Iron box
5. Paper cutter
6. Hammer
   Plastic mug
   Rope for handle
   Clip
   Eyelet
   Gum
   Nylon rope
   Gada cloth piece
   Star and hole punch
   Product Samples.

Raw Material Selection

Used white paper is the best material; glossy papers are not useful for eco-friendly hand made paper making. Glue and staple pins are to be removed from paper before soaking. Tear the paper and soak the pieces overnight in water.

Mould and Deckle

This simple device could be made easily with a carpenter's help. The mould comprises a rectangular wooden frame. A layer of mesh (nylon net) is stretched on this mould.

The deckle (top portion) made of wood, is of the same size as the mould and is an open frame that rests on top of the mould. The mould is to be attached firmly with deckle by hook and eye clips or press-stud. The wet pulp is deposited onto the mesh before it is pressed and dried into sheets of paper.

Prepare the mould and deckle as follows: Take 180 cm (6ft) length of 2cm (¾ in) square wood and cut it into, (i) four 25cm (10 in) lengths and, (ii) four 20cm (8in) lengths; nylon mesh with 12 to 20 holes per cm (50 holes per in); brass or stainless steel pins, or staples; waterproof adhesive; nails. Now arrange the cut wood to make two frames of the same size and shape. The mesh must be stretched tightly over the top of one of the frames. Nylon stretches when it is wet, so dampen the mesh before stretching it. Excess mesh should be trimmed.

Pulping

The waste raw materials soaked in the water is blended in a mixer to get the pulp.

Formation of Paper

The cuddaph or cement tank with one-third water should be kept ready for placing mould and deckle unit. The pulp should be stirred well before placing on the VAT. The mould portion should be submerged in water when pulp is poured. The pulp is to be formed evenly on the VAT; the VAT is firmly held while lifting in order to allow the water to drain back through the mesh. The thickness of the sheet is determined by the quantity of pulp.

Separation of Sheets

Separate mould and deckle by unlocking the hooks. The evenly formed wet sheet/pulp on the mould is removed/separated by placing gada cloth uniformly over the pulp. This process is repeated to make many sheets.

Screw Press

The evenly formed wet sheets on the hand vat are removed by placing gada cloth interfacing over the finished sheet and this process is repeated to make subsequent sheets. This wet paper contains moisture and air bubbles. The water remains in the wet paper has to be removed thoroughly by applying even pressure on the paper so as to obtain smooth surfaced paper. The cloths with wet paper are stacked in the screw press.

Drying and Calendering

The evenly formed wet sheet on gada cloth is allowed to dry by hanging on the ropes with clips. The sheets can be removed from the gada cloth once they are dried. The dried sheets are calendered, with the electric iron. The calendered sheets are then trimmed.

For production of hand made paper from agricultural residues, an investment of Rs.6-7 lakhs on equipment such as beater, agitator, tank, cylinder mould machine, calendering machine and cutting machine is required.

Using the sheets made through the above-mentioned process one can make the following paper products:

Writing papers, Tissue papers, gift bags/box, greeting cards, book covers, paper towels and plates, business cards, wallpapers, birthday cards, files and folders (small size) and envelopes.

Source of Technology: AMM Murugappa Chettian Research Centre, Taramani Chennai-600113.

Modernisation of Bamboo Based Industries

Bamboo has been in use in India for thousands of years. It is well known as enduring, versatile and highly renewable material. Bamboo has been integral part of the cultural, social and economic traditions of rural and tribal societies of the country. Millions still depend on it for their livelihood and for house hold and functional uses. Community has realised the importance to nurture and protect bamboo and is repository of vast knowledge and skills related to the propagation, processing and usage of Bamboo. India is endowed with large resources of bamboo, the second largest in the World. There are 130 species spread across 18 genera. They grow naturally at heights ranging from sea level to 3500 metres. Bamboo grows in almost ten million hectares of forest land. Also it is cultivated in substantial quantities on revenue, private land and as captive cultivation by paper mills. North East accounts for almost two-third of produce of the country. Eighty per cent of the stock in the country comprises of the following species–*Dendrocalamus strictus* (53 per cent), *Banbusa arundinacea*–15 per cent and *Melocanna baecifera* 15 per cent and rest of other varieties. The immense potential of bamboo has been recognised and Govt. of India has set up National Mission on Bamboo Applications–(NMBA) to promote use of this vast resource with skill development and value addition to make it a self sustainable activity. A Cane and Bamboo Technology Centre (CBTC) has also been set up at Guwahati with UNDP assistance to give boost to Bamboo industry in the country.

Bamboo provides the base for a broad range of rural and semi urban cottage industries that provide livelihood for rural masses particularly home based workers comprising men and women in the unorganised sector such as handicrafts, furniture, Agarbattis (incense sticks) which have become a necessary in all religious functions as well as air freshners and mosquito repellent sticks. In addition to domestic market it has international market as well. The current international incense market is valued at US $ 1.0 billion per annum. As such there is very wide market for agarbatti sticks. India is the largest producer of agarbattis in the world. The price range is Rs.30-35 per kg for good quality bamboo sticks–made by using stick making machines. The slivers of uniform size and thickness are fed to stick making machine which produces quality sticks of uniform size.

There is need to set up Common Facility Centre (CFC) where various machines could be set up for providing quality slices of bamboo to bamboo stick making units, matt making units and manufacturers of handicrafts. Entrepreneurs can also set up their mechanised units with financial support of banks.

Central and State Handicraft Boards can also be approached for financial and marketing support.

Several mechanised units for production of bamboo agarbatti sticks are functioning in Assam, Nagaland, Arunachal Pradesh Meghalaya etc. NMBA also provides financial assistance to prospective entrepreneurs. CBTC Guwahati has also developed wood substitutes and a plant is functioning near Guwahati. Advances in structural engineering and the development of bamboo based composites have made possible light weight and aesthetic construction in housing and community buildings.

The Process for Bamboo Mat Making

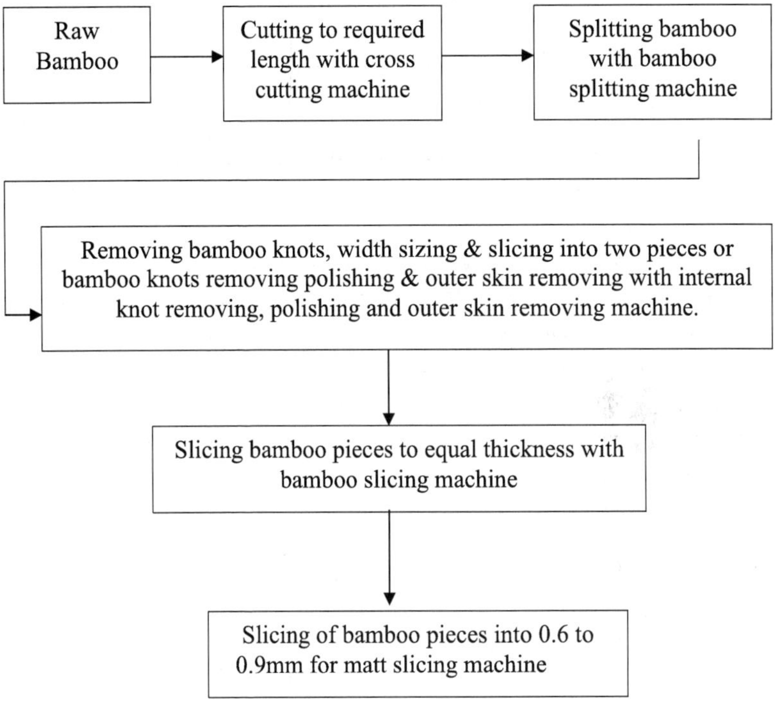

Machines Required for Common Facility Centre

1. Cross cutter
2. Hydraulic splitting machine
3. Inside knot removal machine
4. Outside knot removal machine
5. Slicing machine
6. Stick making machine
7. Stick sizing machine
8. Two side planner
9. Four side planner
10. Sharpening machine.

Manufacturers

1. M/s Sugga Engineering Works (P) Ltd., 58, Rani Jhansi Road, Paharganj, New Delhi–110055
2. M/s Garnet Tools, Industrial Area, Ujjain Road, Dewas M.P.
3. N.S. Machine Tools, Ludhiana.

Products

Bamboo sofa, centre table, easy chairs, weaved bamboo window blinds and screen, fruit forks, party picks, back scratchers, ear scrappers, rice spoons, tooth picks, ice spoons, brooms, handbags, flower vases, ash trays, wall hangings, folding fans, walking sticks, free guards, lamp shades of various design have domestic and internal market.

Central Handicrafts Development Corporation can provide marketing support.

Fish Culture

Fish culture is very popular in states of Andhra Pradesh, West Bengal, Orissa, States of North East, Chhattisgarh etc. where water resources are easily available. Interest in fish culture is increasing to improve quality of life and higher economic return than conventional agriculture. During 1950 fresh water fish production was 30 per cent and rest of the demand was met from sea. With modern scientific interventions fresh water fish production has increased from 30 to 45 per cent where as production from sea has come down from 70 to 55 per cent. In this regard research, development and training facilities were extended by institutes like CIFE, CIFA, CMFRI, CIFT, Madurai Kamraj University and many other such institutions. Apart from research in genetics of fish and prawn they have made biotechnological intervention for protection of environment and abatement of pollution. CIFT has made important contribution for maintaining health standards and quality control for exports of prawn and other marine products. Technologies have been developed for making fish pickle, fish keema ready to cook fish products and for preservation, drying and packaging fish products. CIFE has set up hatcheries in various states for providing fish seedlings to farmers. The tested technology offers high success rate and help in providing large scale self employment and increase production and also help to fight malnutrition amongst women and children. CIFE has trained more than 1100 farmers who are successfully carrying out fish culture. The earning of each family is more than Rs.5000 p.m.

Oxygen plays a vital role in intensive culture of aquatic organism. Aeration helps in reducing the level of biochemical oxygen demand and increasing primary production. Oxygen eliminates anaerobic conditions in the culture system. CIFE developed airlift incubatory system as early as 1978. Since then mass culture of Zooplankton such as Artemia, Branchionus sp. has been demonstrated. Use of compressed air for hatching of carp eggs and rearing of fry ensure 95 per cent survival. In Hi-tech system, 5000 spawn can be raised in 1 cu.m. water.

In pisciculture the present production of 1.5-2.0 ton/ha/year can be increased to 7-8 ton/ha/year. Keeping the selling price as low as Rs. 25-30/kg. farmers can earn profit of Rs.1.00-1.50 lakh/ha/year after deducting cost of feed and maintenance. CIFE has also developed model of mixed farming by including horticulture, pulse growing, dairy etc. Entrepreneurs should contact Central Institute for Fisheries Education, (CIFF) Versova, Mumbai for Technology and Training so as to make use of hi-Technology. Funding can be obtained from banks, NABARD etc. State Fisheries Department should be contacted for local support and help.

Proper feed is very essential for growth of fish. Feed plants have been set up at Central Institute of Fisheries Advancement (CIFA) Bhubaneswar and Central Institute of Agricultural Engineering Bhopal. CIFE has also set up fish feed mill at Baba Amte Complex. Pelletized feed enhances production and reduces waste which pollutes water. Aeration, use of blowers and agitators improve oxygenation which also improved yield of fish.

Source of Technology: Director CIFE Mumbai/Dr. S.N.Dwivedi, Former DG, MAPCOST Bhopal

Agarbatti Manufacturing

Burning of Agarbbaties during prayers at temples, Mosques, Gurudwaras and other religious places has become a ritual to create congenial atmosphere and purify the environment. The use of agarbattis is of recent origin, around middle of nineteenth century in North India although the art of manufacturing agarbattis was developed at Tanjore and adjoining areas during the rule of Marathas but it progressed during erstwhile state of Mysore due to patronage it received from rulers. Now Bangalore and Mysore are major centers of Industry in India although there are clusters manufacturing agarbattis in other states. The word agarbatti was coined perhaps due to use of powder of Agarwood as one of the ingredients in its manufacture. The Indian agarbatti industry has made significant progress and its exports are $ 1.00 billion and at least Rs. 5000 crore production is consumed in India.

Agarbbatti making is labour intensive cottage industry and employs few lakh persons particularly women. Although agarbatti machines have been developed and some automatic units have been set up in Karnataka but it is still a cottage scale industry and has potential to generate employment opportunities in rural areas.

In good old days natural materials like powders of Sandal wood, agar wood, Deodar wood, roots, barks, seeds, leaves, flowers, gums and resins in addition to charcoal, saltpetre, colours, wood gums like 'Maida lakdi' were used.

Process of Manufacture

The process is generally carried out manually, especially by women which involves following steps:

1. Processing of raw materials.
2. Agarbatti rolling.
3. Perfuming and packing.

Processing of Raw Materials

Various raw materials like Rostun, Talispatre, Kachram, Kasturi-gadde, Nagarmothes, Khus roots, Davana etc. are mixed and powdered in a disintegrator or pulveriser, subsequently gums like Maida Lakdi, Guggula, aromatic chemicals, perfumery compounds and charcoal are added, and ground into fine paste after mixing adequate quantity of water. Manual grinder could be used for grinding of raw materials. Some traders also supply the ready mix to domestic agarbatti makers.

Agarbatti Rolling

The mix (paste) is rolled over bamboo sticks on smooth wooden planks by hands. It is a skilled job and needs few days training. After rolling the sticks are cut in uniform size. During rolling colours, sandal wood dust etc. are added to give them fine finish.

Perfuming and Packing

The rolled sticks are dried in shade and perfumed by dipping in solution of aromatic chemicals. Now a days, synthetic chemicals are used which are available at cheaper price for supply in competitive market. But for superior products natural essential oils and perfumes are used. Finally, the bundled sticks are packed in cellophane paper to preserve the aroma and card board boxes.

Fragrance and flavour development Centre (FFDC) Kannauj U.P. a Ministry of Industry (G.O.I.) organisation, offers specialised training in the production of agarbattis of high quality.

Instant Soil pH Meter (Developed by CSIO Chandigarh)

Measurement of pH of agricultural soil is very essential to assess the nature of soil. The pH is determined either by measurement of the electro chemical potential that exists between different solutions or by an electrode system, which is sensitive to the presence of hydrogen ions. The potential developed between two electrodes is measured. The pH meter consists of two metallic electrodes for sensing and reference. The measured values are readable directly on an analog meter.

Broad Specifications

Range: 3-11 pH

Accuracy: 0.2 pH

Sensor: Metallic

Salient Features

- Direct measurement of pH *in situ* sal.
- Unbreakable metallic electrodes.
- No batteries to be replaced.
- Portable, rugged instrument.

Application

- ✰ Farmers
- ✰ Agricultural Universities
- ✰ Soil Testing Labs
- ✰ Tea gardens, Nurseries, Fisheries etc.

Some Post Harvest Technologies Developed by CIPHET

Central Institute of Post Harvest Engineering and Technology (CIPHET) Ludhiana (Pb) has developed number of tools, gadgets and machinery to pre-process fruits and vegetables.

CIPHET has developed Maturity tester for mangoes on tree or at the storage godowns which predict maturity on the principle of colourimeter. Change of peel colour and total soluble solids are indication parameters to measure maturity of mango. The technique can be employed to sort out the mangoes at export posts, big mandies and processing plants.

Harvesting tools developed are strawbury clipper and Banana comb cutter. It is useful to de-stem the comb from banana bunch. It replaces use of sickle which is labour and time consuming. It is safe for human beings and works faster.

In CIPHET Tomato Grader, the tomatoes slide down the pipes due to gravity and fall immediately wherever they find the space depending on their size. The pipes are inclined at 10° so that the tomatoes slides directly in crates. The gap between pipes and inclination could be adjusted. It can also be used for other round shaped fruits and vegetable.

The Institute has also developed ground nut Pod Grader and Decorticator. The unit consists of feed hopper, decorticating drum, aspirator and sieve grader to grade kernel in two grades. The capacity of unit is 60 kg/hr and price is approx. Rs.60,000/-. The dehulling efficiency is 87 per cent with no breakage of Kernel.

Lac scrapper cum graders of two capacities to scrap 3-4 lac sticks to 10 sticks and simultaneously crush and grade the scrapped lac in three grades. The capacity is 20kg of lac sticks/hr for lac sticks of kusum plant at a moisture content of 3 per cent (d.b) after two passages, through this machine.

CIPHET has also developed Rotary Maize Cob sheller which is suitable for all varieties with almost zero damage to grain. Capacity is 65kg/hr.

The Institute has also developed rural cold storage using porous bricks developed by them. One can have a look at all these and many more devices at the Institute.

Sisal (*Agave Sislana*)

Sisal is a hardy plant, grows in arid and semi-arid tropics with moderate humidity in almost all types of soils. Sisal plantations are found abundantly in various parts of India on the sides of Highways and field bunds. They conserve the soil by preventing run off. The plant grows from Central bud with long (1-2m), erect and thick flesh dark green leaves. For cultivation plant to plant space is kept 1.2-1.8 m. A single plant yields 250-300 leaves in life span of 8-10 years. The width of leaves vary from 7-8 cm at the base to 10-15 cm in the middle.

Composition of Leaf

Yield (1000 kg Sisal leaves)

Fibre: 4 per cent	Fibre: 30 kg.
Cuticle: 1 per cent	1 kg Hycogenin: 1 kg. (a steriod)
Moisture: 87 per cent	Wax: 3.75 kg.
Other dry matter: 8 per cent	Paper: 33 kg.

The fibre is bio-degradable and eco friendly.

Processing

The processing does not involve any health hazards and following three methods are involved:

1. *Retting*: Cut the leaves, dip in water till fibre separates out, beat to remove pitty matter, wash and sun dry.
2. *Boiling*: Boil the leaves, subsequently beat, wash to extract fibre and sun dry. It is faster than retting.
3. *Mechanical*: A machine called Raspador is used to separate fibre from leaves which are inserted in the machine and fibre is pulled out. This method is quick and does not require any pre-processing and is cost effective.

Chemical Composition of Sisal Fibre

Cellulose	66-72 per cent
Hemi-cellulose	12 per cent
Lignin	10-14 per cent
Pectin	0-9 per cent
Moisture	6-10 per cent

Structural Properties of Sisal Fibre

Length	0.60-1.20 m
Diameter	0.05-0.20 mm
Tensile strength	500-660 mN/m^2
Colour	Creamy white.

Applications

Sisal fibre has been in use for making ropes, twines etc. It has been put to other uses like Handicrafts, Textiles and upholstery materials, matts, matting bags, hand made paper industry, fibre composites (in place of glass fibre), automobiles, Building materials etc.

The residue obtained during fibre extraction is used in Vermi-composting which offers additional income to the entrepreneur.

Source of Technology: Advance Materials and Processes Research Institute (earlier RRL), Bhopal.

Vegetable Tanning of Leathers

Improved Rural Tanning of Vegetable Tanned Leathers.

Introduction

In India Rural Leather tanning has played prominent role for many centuries since it has benefited rural flayers, footwear and leather products manufacturing by village artisans from the tanned leathers obtained from village tanners. They catered to the day to day requirements of Farmers and others residing in villages, and small towns. It helps substantially in the rural economy.

Rural tanners besides procuring hides and skins from native flayers, used to collect barks, shrubs, and fruits for vegetable tanning either free of cost from nearby forest area, trees near rivers or ponds or by paying nominal charges to its suppliers and lime was also obtained from local kilns. Thus rural tanning was fully integrated and self sustainable economy.

This system suffered in course of time as the finished leathers produced used to emit unpleasant odour, and became hard due to incomplete tanning and did not fulfil the needs of sophisticated customers. Moreover, the tools and techniques became old and full of drudgery which forced out many artisans from the profession.

This necessitated development of improved processes for rural tanning through constant endeavors of institutions like Central Leather Research Institute, Chennai and its Regional Extension centers located in different zones of India and many State Government organizations, N.G.O. and Khadi and village Industries institutions.

It is worth mentioning that vegetable tanning is Eco-friendly, discharging minimum pollutants and waste products such as bark, nuts, fruits which are a good source of organic manure for use in Agriculture.

An improved and modified process of rural vegetable tanning developed at Central Leather Research Institute and demonstrated at many rural leather clusters is given below:

Raw Material

Locally procured green, semidried or salted hides and skins are used for vegetable tanning. Generally, buffalo and cattle hides are processed into tanned leather. Lower grade of hides and skins are preferred and good quality hides are sold to hide and skin traders from cities and towns as they fetch remunerative prices and lower grade hides give good returns after conversion into vegetable Tanned leather. Skins of sheeps and goats are commonly not processed by rural tanners since their demand for manufacturing leather products is not attractive. However in rural areas around Kolhapur and Sholapur in Maharashtra vegetable tanned goat skins have good demand for making Kolhapuri Chappals.

Soaking

Hides and skins are soaked either in pit of dimensions 6' x 4'x4' constructed above the ground with suitable outlet of waste water into an underground drain.

While green (fresh) hides require three to four hours time for soaking but in case of salted hides soaking is carried overnight till hides become soft and pliable. Dried hides in most cases are soaked for 18 to 24 hours depending upon the condition of hides. Prior to commencement of soaking hides and skins are weighed in order to work out the requirement of chemicals and Tanning materials. However, due to any reason if hides are not weighed prior to soaking this should be followed after soaking and draining excess water from hides to record soaked weight as it helps in determining the quantity of chemicals and tanning materials in the subsequent process. Nearly 500 grams of bleaching powder and 500 grams of washing soda (sodium carbonate powder) may be used for a pack of 18 to 20 dried hides to prevent any put-rification during soaking as increase in time at this juncture leads to bacterial damage. It is better to wash hides in fresh water prior to actual soaking to remove adhering dirt, blood, and other impurities from hides. Sometimes, one more wash is given after first soaking for about 2 to 3 hours so as to ensure sufficient soaking followed by liming. It may be noted that proper soaking and liming determines the quality of vegetable tanned leathers specially from rural leather clusters.

As far as possible hides should be flat and quite soft and slightly plumped after soaking.

Liming

Bigger hides are cut into equal half from head to tail along backbone line for ease of processing and handling by local artisans.

The soaked hides or skins are immersed in first pit containing a solution of 200 per cent old lime liquor (once used) 200 per cent fresh water and lime powder (calcium hydroxide) and 1 per cent sodium sulphide. The dimensions of lining pits are same as that of soaking pit. The liming period normally ranges from 6 to 7 days and it can be curtailed by one day depending upon the condition of soaked hides.

Hides ae daily hauled and replaced from first pit also called old liming pit daily two times in the forenoon and evening and hauling out side for about 30 minutes. This hauling and replacing helps in agitating lime liquor and early loosing of layer from grain side and developing mild swelling and plumping. After keeping in the old lime liquor pit hides are unhaired with a curved semi blunt knife to scrap of the hair and epidermal layer from the grain surface. Hides are now taken for reliming.

Reliming

Duration 3 days.

Following ingredients are added to the pit water (sufficient water to keep the hides well immersed in lime solution avoiding any sort of oxidation)

Lime powder	6-8 per cent
Sodium Sulphide	1 per cent
Caustic Soda (Sodium Hydroxide)	0.5 per cent

Similar to old lime liquor, here too hides/skins are hauled twice daily for 3 days. During this period the flesh swells sufficiently and hides are ready for fleshing.

Hides on seventh day morning are lightly rinsed in the fresh water and taken for fleshing.

For efficient and quality production it is preferable that hides now called pelt are fleshed on a wooden beam kept in inclined position. Fleshing is done using a semi curved sharp single edge knife. Some rural tanners prefer to use RAMP for fleshing and carrying out fleshing on a flat big smooth stone slab but it results poor output and causes back pain to artisans due to excessive bending of body.

Rubber gloves must be used during fleshing work to avoid contact of hand and palm with limed pelt containing injurious chemicals like sodium sulphide and caustic soda. After fleshing pelts are scudded on grain side with semi curved blunt knife to remove hair roots, fat, etc.

Fleshed weight of pelts is noted down in order to calculate the quantity of chemicals in subsequent operations.

Fleshed pelts are washed in fresh soft water to remove surface lime, sulphide, scudd etc.

Deliming

Deliming is carried out in a pit for overnight using following chemicals:

Water	100 per cent
Ammonium sulphate	1 per cent
Sodium Bisulphite	0.5 per cent

Next day pelts are trampled in deliming bath for about one hour with feet to remove the adhered lime.

Completion of deliming can be checked with phenopthalein solution. Pour few drops of phenophthalein on cut section of pelt. If it show a streak of pink colour it indicates that deliming is almost complete. If phenophthalein is not available then deliming may also be checked with Turmeric powder solution. Pour few drops on cut section of pelt. A streak of pink colour indicates completion of deliming. After deliming pelts are once again scudded with blunt semi curved knife by placing pelt on inclined wooden plank, tree or locally available knot free beam. This is done to remove the remaining scudd from the pelt.

Pelts are washed again in fresh soft water and water is drained out for 8th day.

The delimed pelts are transferred one by one to the salt solution made earlier in a pit

Common salt	5 to 6 per cent
Water	50 to 60 per cent

and kept here for about an hour.

Later diluted sulphuric acid solution prepared earlier is added very slowly to the pit containing salt solution.

Sulphuric Acid	0.5 per cent	Pelts kept in the
Water	10 per cent	pit for about 2 hours.

Note: Percentage of chemicals used above are based on pelt weight.

After further handling the pelts in the pit, pH of the cross section of a piece of pelt cut from pelt is checked with the help of pH paper and when pH reaches to 5.5 (Approx). Pretanning is carried out in the same pit so as to facilitate uniform tanning at later stage.

Pretanning

1 per cent of pretanning syntan such as Basyntan P of BASF or P vernatan P of colourchem or any suitable pretanning syntan and 10 per cent water

Syntan is dissolved in water and added in 3 to 4 instalments at an interval of 30 minutes. After the addition of last instalment, pelts are further handled with hand in this pit and left immersed overnight.

Malani

9th day.

Malani which is also called colouring pit is the beginning of first step of prior to bag tanning. In one vegetable tanning pit of dimensions 6' x 4' x 4' which is already in use the following materials are added:

Mixture of Babool Bark and crushed myrobalan nuts in the ratio of 3:1.

25 per cent of the above mixture is added to a vegetable tanning pit containing 300 per cent once used tanliquor. This pit is kept ready one day earlier prior to Malani process. The strength of vegetable tan liquor in this pit should be about 15 degree barkometer. The leathers are placed one by one in this pit and handled with hands continuously for 3 to 4 hours to avoid any sort of oxidation and development of patches so as to obtain uniform colour on the grain surface. In the evening again after handling for one hour leathers are well immersed in the pit avoiding air pockets between leathers. It may be noted that in the absence of proper handling in the malani tanning pit chances of patches on grain surface of leather are quite possible.

Leathers are kept in the malani first tanning pit for 2 to 3 days depending upon the convenience of vegetable tanning.

12th Day

Bag Tanning: Actual bag tanning consists of two parts. In the first part, the leathers undergone preliminary or partial tanning are stitched using locally available date or palm leaves or sisal fibre or moonj. The bag is stitched from butt region of the leather to neck region. The bag of leather stitched on sides is filled with babool bark and crushed myrobalan nut mixture in ratio of 3:1, leaving a wide opening near the neck region to pour tan liquor. For three days the tan liquor is poured at regular interval in the leather bag placed over a wooden log. The leather bag is suspended from a wooden log or pole kept prependicular over tanning pit of size 7' x 4' x 3' filled with vegetable tan liquor. For three days pouring of vegetable tan liquor is continued avoiding exposure to direct sunlight. By this time the butt region and middle region along back bone gets tanned. The completion of tanning in thickest portion of butt region is checked by knocking with hand giving typical sound.

15th Day

On 15th day the bag is lifted and removed from the wooden pole kept on the base of bag tanning pit on a used gunny bag and neck side opening is stitched with palm or date leaves or moonj and a wide opening is made with the help of sharp steel knife on the base or tip of the butt portion for filling with tan liquor. The leather bag is again hanged on the wooden pole kept over bag tanning pit and tan liquor pouring is continued for 2 days to complete the vegetable tanning in the neck region. After ensuring complete vegetable tanning the bag is removed from the wooden pole cut open to remove tanning materials filled on the flesh side of leather. The used bark is transferred to the bag tanning pit to extract remaining tannin content.

The bag tanned leather is lightly rinsed in the tan liquor available in the bag tanning pit.

17th Day

If necessary bag tanned leather may be bleached to remove any undesirable stains on grain side and further improve the colour into light cream colour with the following chemicals.

Sodium Bisulphite	1 per cent
Water	20 per cent

Applied with a used gunny cloth or salt goat hair brush on grain side and kept in this condition for 30 minutes.

Next apply solution of:

Oxalic Acid	0.5 per cent
Bleaching syntan	0.5 per cent
(Like FCBJ-3 (BASF) if available)	
Water	20 per cent

This mixture is gently applied on grain surface of tanned leather and kept for 45 minutes to 60 minutes. After this the leathers are lightly rinsed in the tan liquor pit.

18th Day

The tanned leathers are some times kept for 2 days in the bag tanning pit layering with 20 per cent vegetable tanning mixture (Equal parts of babool bark and crushed myrobdan nuts). This tanning mixture is sprinkled on flesh side of leather keeping in pile flesh to flesh and well immersed in the tanning pit. This layering with legetable tanning mixture further improves vegetable tanning.

20th Day

Leathers are taken out from vegetable tanning pit rinsing well to remove adhering bark mixture.

The tanned leathers are kept in a pile for 3 to 4 hours either on a flat wooden platform or over a R.C.C. table of size 7' x 4' x 3', having sinch thick top to drain excess of tan liquor.

18th to 20 Day

The tanned leathers are hung in a row close to each other tied by jute ropes inserted in the small holes made on tips of the butt and neck region for conditioning. The room in which vegetable tanned leather are hung may have windows near to roof but there should not be exposure to direct sunlight as sunlight leads to oxidation and darkening of colour of tanned leather. After ensuring the proper conditioning (removal of excess moisture), the leathers are removed from the wooden poles by untying the knots.

Setting

20 day: Leather are hand set on the R.C.C. table described as above with the help of wooden semi curved blunt stone or stainless slicker to remove wrinkles and making the grain surface smooth and leathers flat.

20th to 21st Day

Hand set leathers are again hung in drying room tied with jute ropes and when almost dry are once again hand set to enhance smoothness, flatness and look of the leathers and hung up for complete drying in the drying room. When completely dry the tanned leather are kept in a pile for one to two days for aging.

Yield

Dried and aged vegetable tanned leathers are weighed to find out the yield-Percentage. Yield calculated on pelt weight basis may range between 42 to 46 per cent depending on the type of raw material and the method of vegetable tanning.

Vegetable tanned leather may be sold on piece basis or weight basis depending upon market practices.

In this way good quality vegetable tanning may be completed in 21 days or alternately vegetable tanning in pits may be carried out using 6 tanning pits starting from 10° barkometer to 50° barkometer gradually increasing strength of tan liquor using babool bark and crushed myrobdan nuts fortified with vegetable tanning extract procured from tanning chemical suppliers. This is known as pit tanning. (Courtesy Sh K.B. Gupta, ex. scientist CLRI).

Appendix

Some Food Processing Machinery Manufacturers

Mysore Precision Engineers C-123, Ind. Est. Yadav Giri, Mysore-570 020	Fruit and Vegetable Processing equipment.
Nalanda Agro-works, Nalanda Nagar, Kurji Patna–800 010	Mini wheat and Dall Mill, papad, press, maize mill, roaster, leaf cup machine, graders etc.
Navin Ajmani, 394, Kohat Enclave Pitampura, New Delhi-100 034	Fruit juice extraction
RPM Engineers (India) Ltd. 14, NP Developed Plots Thiru Vika Ind. Estate, Ekkathuthangal, Chennai–97	Small Scale Milk Processing Plant, Ice cream Plants.

Rajalakshmi Industrial Agencies 30/1, Silver Jubilee Park Road, P.B. No. 6690, Bangalore–560 006	Wet grinders Potato slicer, peeler
Raylon Metal Works Ramakrishna Mandir Road Kondivittavillage, Opp. Marol Bazar, J.B. Nagar, Andheri (E), Mumbai–400 059	Fruit and Vegetable Processing Eqpt.
Shnmugham Pillai and Sons 12, Chelladiamma Koil Street Dindigal–642 010	Continuous Roaster
Siddivinayaka Machinery and Equipment No.10, 3rd Stage, Industrial Suburb, Mysore–570 008	Cleaners, graders, driers, destoners, air classifiers, roasters roaster, mini dall mill, flour mill dehuskers, oil extraction.
Techno-equipments Gali No.9, Sathyam Ind. Estate Off. Bhakta, Karvi Devshi Marg, Behind V.S. Vitamin and Pharma Corporation Ltd. Govandi, Mumbai–400 008	Fruit and vegetale process plant.
Dhopeshwar Engineering Pvt. Ltd. A-16, Co-op. Estate Balanagar, Hyderaad–500 037	Processing Plants for Aloe- vera, Biodiesel, Aromatic oils.
Central Institute of Agricultural Engineering (CIAE), Nabi Bagh, Berasia Road, Bhopal.	Potato Slicer, Soyabean dehusker dall mill etc.
Trinitron Products S.No. 58/1/1, Shri Harinarayan Complex, lock No. A/9, Vadagaon Block Off. Mumbai- Bangalore Highway, Pune–411 041	Spray driers, fluid bed driers, tray driers.

Baba Industries H-Block Ind. Area Govindpura, Bhopal.	Honey processing plant (Small)
Wood Fabricators Shop No.3, Avanthi Appts. Lakshman Mhatre Road, Navagaon, Dahisar (West) Mumbai–400 068	Wooden Vats, Filter presses and auxiliary equipment.
Vijaya Industrial and Enggs. Works, 45/4, 7th Cross 80 Ft. Road, 6th Block, Rajaji Nagar, Bangalore–560 010	S.S. Tank, Vessels, Heat Exchangers, Kettles.
Bharat Engineering Works, Budhwaria Bus Stand, Ujjain, M.P.	Leaf/cup plate making machine, papad press, potato chiping m/c, Broom making machine.
Swaraj Fabricators, Faizabad Road Bara Banki, U.P.	Distillation Plants for Aromatic Oils, Spice oils, Baby Boilers.
Varsha Boilers 505, Churchgate Chamber 5 New Marine Lines Mumbai–400 020	Steam generators.
Thermax (I) P. Ltd. Plot No.7, D-1, Block MIPC, Industrial Area, Chinchwad, Poona–411009	Boilers.
Gardners Corporation 6, Doctor Lane, Near Gole Market, P.B. No.299, New Delhi–110 001	Fruit and Vegetable processing equipment.

Grovers Pvt. Ltd. 4-F-4, Shankardham, Sundervan Complex, Off. Lokhandwala Complex Road, Andheri (W), Mumai–400 008	Dehydration plants.
International Food Machinery Opp. Deep Bhawan Pandit Nehru Marg, Jamnagar–361 083	Dehydration machinery Blancher, Hammer mill.
M. Sant Singh and Sons 3794, Morigate, Delhi–110 006	Beverage machinery
Mangal Engineering Works Factory Area Patiala–147001 Punjab	Bakery confectionary plant, pharma machinery
Mysore Industries P.O. Box 310, Shed 57/1, 2nd Stage, Industrial Suburb Mysore–570 008	Grain processing machinery Mini flour mill
B. Sen Barry and Co. 65/11, New Rohtak Road, New Delhi–110 003	Pulper, fruits and vegetable processing equipment.
Baker Enterprises 23, Bhera Enclave, Near Peera Garhi New Delhi–110087	Bakery machinery
Batlibai and Co. P. Ltd. 98/A, Armanium street Chennai–600 001	Pulverisers, Hammer Mills.
C.G. Dandekar Machineries Dandekar Wadi, Bhiwandi–421 302 Thane, Maharashtra	Rice mill machinery, cleaners, shellers, polishers, separation equip.

D.P. Pulverisers Modi and Modi Building 2 76, Nagindas Master Road Fort, Mumbai–400 023	Pulverisers
Dhiman Sales Corporation Dakhni Gate, Nakodar-141310 Dist. Jalandhar	Confectionery machinery
Edward and Sons Ambedkar Nagar, Deevarajeevan Halli, Bangalore–560 045	Bakery equipment.
Durai Industrial Works, 1143, Mettupalayam Road, Coimbatore.	Poha making machinery, paddy roaster, pre-cleaner edge runners.

Index

MINI RICE MILL

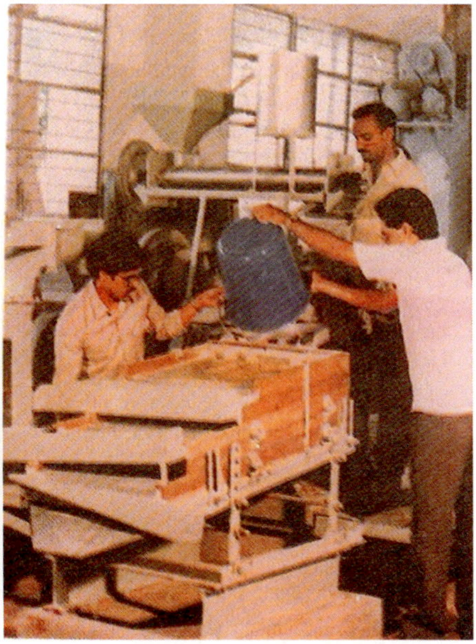

MINI FLOUR MILL

IMPROVED DRYING OF RED CHILLIES

IMPROVED KILN

JAPANESE QUAIL PRODUCTION

LEAP CUP MACHINE

PADDLE OPERATED POTTERS WHEEL

IMPROVED GUR FURNACE AND CHIMNEY

DISTILLATION PLANT FOR ESSENTIAL OILS

BUILDING MATERIAL

C-BRICK MACHINE

CONCRETE BLOCK MACHINE

SOLID/HOLLOW CONCRETE BLOCK MACHINE

RURAL TOILET

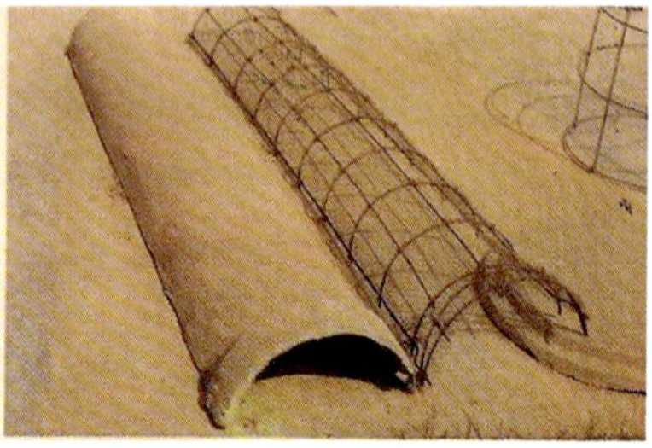

FERRO-CEMENT CHANNEL

FOOD PROCESSING

MINI DALL MILL

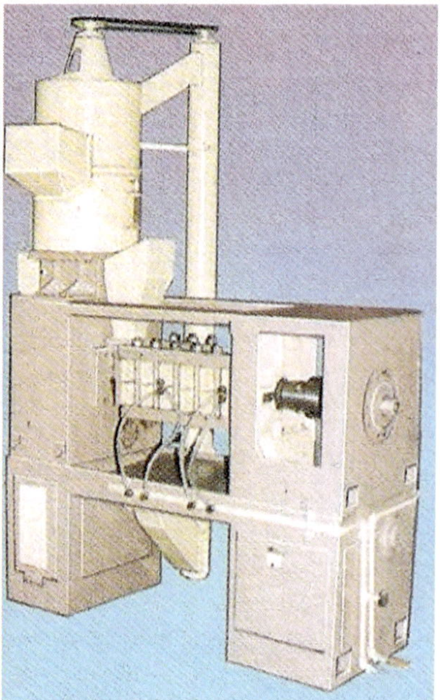

MINI OIL EXPELLER

ENERGY

SOLAR WATER HEATER

SOLAR PANEL

SOLAR LANTERN

WIND MILL FOR
POWER GEN

GASIFER PLNT

MEDICINAL AND AROMATIC PLANTS

CULTIVATION OF LEMON GRASS

JAPANESE MINT

AGRICULTURE MACHINERY

NAVEEN SICKLE

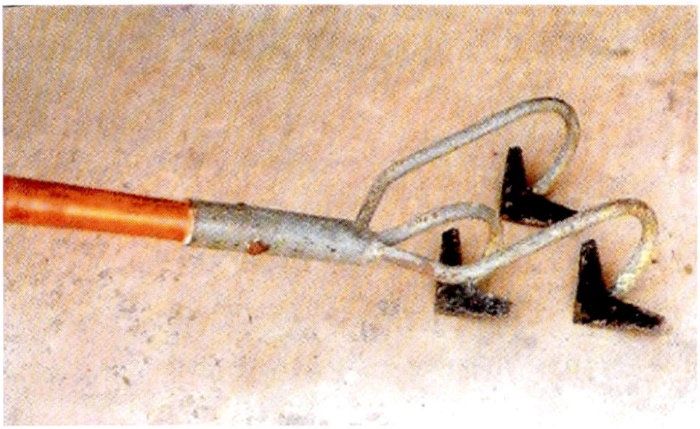

GRUBBER WEEDER

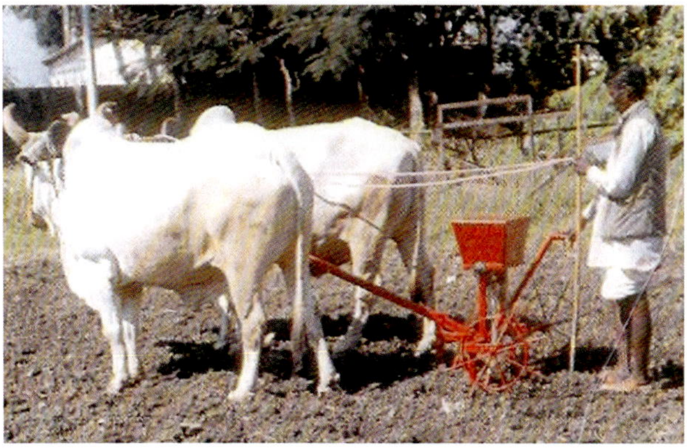

LOW COST SEED DRILL

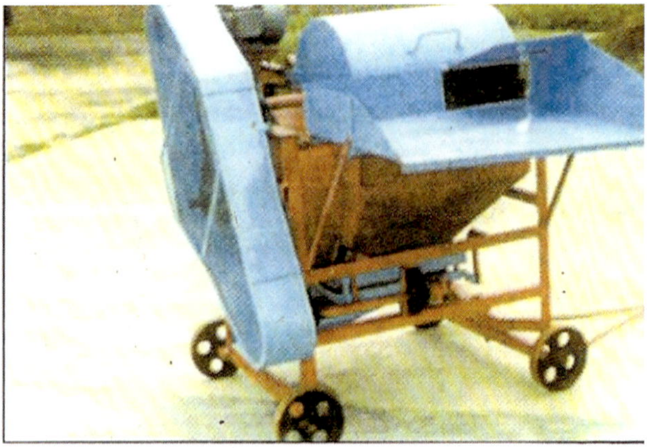

MULTICROP THRESHER

IMPROVED BAKHAR

POWER OPERATED TILLER